Jürgen Mittelstrass

Theoria

Jürgen Mittelstrass

Theoria

Chapters in the Philosophy of Science

DE GRUYTER

ISBN 978-3-11-059593-2
e-ISBN (PDF) 978-3-11-059668-7
e-ISBN (EPUB) 978-3-11-059282-5

Library of Congress Control Number: 2018941016

Bibliographic information published by the Deutsche Nationalbibliothek
The Deutsche Nationalbibliothek lists this publication in the Deutsche Nationalbibliografie; detailed bibliographic data are available on the Internet at http://dnb.dnb.de.

Printing: CPI books GmbH, Leck

www.degruyter.com

In memory of Gilbert Ryle and Alistair C. Crombie

Contents

Preface XI

I Concepts

1 Complexity, Reductionism, and Holism 3

2 Predictability, Determinism, and Emergence 12

3 Discovery 22

4 Time 27
4.1 Arrows of time 28
4.2 Time in the lifeworld 36

5 Limits of Science? 41
5.1 Completed science? 42
5.2 The sphere of knowledge 45
5.3 Limits of science 46
5.4 Productive incompleteness 53

6 Transdisciplinarity 57
6.1 Disciplinarity, interdisciplinarity, and the new complexity of science 57
6.2 Transdisciplinarity 59
6.3 The Unity of Nature 62
6.3.1 Nanotechnology 62
6.3.2 The quantum-mechanic measurement process and the concept of information 63
6.4 Methodical transdisciplinarity 65

7 Pragmatic Dualism in the Philosophy of Mind 68
7.1 The *ego* 69
7.2 Science and the philosophy of mind 71
7.3 Free will 74

II Nature and History of Science

8 From Plato's World to Einstein's World 79

9 Causality in Greek Thought 87
9.1 Becoming and passing away 88
9.2 Platonic causes 89
9.3 Aristotelian causes 93

10 Scientific Truth, Copernicus, and the Case of an Unwelcome Preface 99

11 Newton's Concept of Hypothesis and the Origin of Empiricism in Physics 105
11.1 Research about hypotheses 105
11.2 Methodological background 109
11.3 Observation and experiment 113
11.4 Analytical and synthetical method 115
11.5 Empirism misunderstood 118

12 Philosophical Foundations of Science in the 20th Century 121

13 The Scientific Mind. Does Science Make Its Own History? 130
13.1 Justified history and factual history 130
13.2 Influencing 136
13.3 Basics and applications 139
13.4 Science as subject? 140

III Ethical and Institutional Matters

14 The Moral Substance of Science 147
14.1 Science as idea 147
14.2 The measure of progress 149
14.3 Ethos in the sciences 152

15 Science and Culture 155

16 Naturalness and Directing Human Evolution 163
16.1 The natural and the artificial 164

16.2 Homo faber ... 167
16.3 The ethical question ... 168
16.4 Concluding remark ... 170

17 Through a Glass Darkly. On the Enigmatic Nature of Science ... 173

18 Quality Assessment in Higher Education Institutions – from the Perspective of Those Assessed ... 176
18.1 First remark ... 176
18.2 Second remark ... 177
18.3 Third remark ... 178
18.4 Fourth remark ... 178
18.5 Fifth remark ... 180
18.6 Sixth remark ... 180
18.7 Conclusion ... 181

19 The Joy and Woe of Scientific Policy Advice ... 182

20 Science – the Last Adventure ... 187

Acknowledgements ... 190

Index of Names ... 191

Index of Subjects ... 195

Preface

Scientific knowledge is characterized by methodical procedures and justification. It is achieved by research and theory formation. But what is a methodical procedure and what are methodically established justifications? What kind of principles must be observed in order to obtain the degree of objectivity that is generally claimed by science? What is the relation between science in the research mode, i.e. in its practical form, and science in the presentation mode, i.e. in its theoretical form? Do the same principles hold here? And how are they justified? Is it even possible to speak of justification in a theoretical sense? Or do we have to be content with less – with corroboration and confirmation? Is the distinction between the context of discovery and the context of justification the last word in methodical and theoretical matters? And how does this distinction relate to that between research and presentation – the constitution of (scientific) objects on the one hand and (theoretical) propositions about them on the other?

These questions, and many others, occur when we reflect on science, when we pursue the philosophy of science. Philosophy of science presents its essential problems and attempted solutions under the headings: theory structure, theory dynamics, and theory explication. Under *theory structure*, philosophy of science deals among other things with the structure and the construction of scientific languages and the structure of scientific explanations and laws. In *theory dynamics* the topics are the reconstruction of scientific developments, the problem of the semantic reducibility of one theory to another and the existence of trans-theoretical criteria for comparing theories with one another with regard to their efficacy. In *theory explication* philosophy of science takes up the analysis and examination of concrete theories, for example, the question whether a theory of space, after the establishment of the definition of congruence, can be definitely determined empirically. Other topics can be added, like *science assessment*, i.e. the establishment of criteria for good scientific research and practice, and *science ethics*, i.e. the reflection on the moral and ethical problems that science is confronted with in the process of research and in view of the applications of scientific results. Is there such a thing as a specific ethics for science? And what is meant by the demand for an *ethos* of the scientist?

The analyses and constructions in this book take up these questions. They are explicitly intended as *philosophical* contributions, not only in the sense implied by the disciplinary use of the term *philosophy of science*, but also in the sense of a reflection on science that, alongside more technical aspects of methodologies and elements of theories, also has an eye for anthropological and cultural aspects.

The first part deals with conceptual and methodical themes. Science expresses itself in concepts and theories; concepts and theories determine scientific developments. Philosophy of science investigates these relations and elucidates them for a general understanding of the world and for science itself, which is often unaware of its own conceptual and theoretical form – as John Locke refering to Isaac Newton put it: "clearing the ground a little" (*An Essay Concerning Human Understanding* [1690], "The Epistle to the Reader"). This is the study of the epistemological foundations of science and of scientific understanding. It includes questions concerning the range of these foundations, i.e. their possible limits. The book's second part opens a historical dimension. By means of examples, a focus is placed on the systematic (but also institutional) relations between history and philosophy of science. The third part deals with ethical and anthropological aspects of science. Science has not only a theoretical but also a practical form. As a rule philosophy of science deals with theoretical aspects of science, but here practical aspects, i.e. ethical and anthropological issues, are emphasized: Science taken not as a particular form of knowledge, but as an institution, as a social organization that is governed by ethical and social norms. The final part continues these institutional reflections on more specific topics like science assessment and concludes with an optimistic outlook: science – the last adventure.

Some of these considerations are based on already published material (mostly in places somewhat off the beaten path.) I owe significant impulses and advice to my former colleagues Martin Carrier, Stephan Hartmann and Peter McLaughlin. The book is dedicated to the memory of my former teachers at Oxford in 1962, Gilbert Ryle and Alistair C. Crombie.

Konstanz, Spring 2018 Jürgen Mittelstrass

I Concepts

1 Complexity, Reductionism, and Holism

There are concepts that belong to the basic terminology of science but which are not used in everyday scientific work – such as the concepts of natural law and causality. Such concepts touch on the epistemological foundations of science, and thus transcend individual disciplines and presuppose a particular interest, the interest in foundational questions of science, and presumably also special skills and competence. Not everything that belongs to these foundations is self-evident and not everything that is said about them in philosophy of science is universally accepted – which in turn lies in the fact that we are dealing with different theoretical approaches. Theory meets theory, and this does not always go without conflict.

In the following, as an introduction to considerations of a theoretical, methodological and epistemological nature, which especially deal with aspects of complex structures, some brief explications of a conceptual nature oriented towards the concepts of complexity, reduction and holism.

1. In a comprehensive presentation of the role that the concept of *complexity* plays in the development of modern science we read: "Complexity determines the spirit of twenty-first century science. The expansion of the universe, the evolution of life, and the globalization of human economies and societies all involve phase transitions of complex dynamical systems."[1] And further: "The theory of nonlinear complex systems has become a successful problem solving approach in the natural sciences – from laser physics, quantum chaos, and meteorology to molecular modelling in chemistry and computer-assisted simulations of cellular growth in biology. On the other hand, the social sciences are recognizing that the main problems of mankind are global, complex, nonlinear, and often random, too. Local changes in the ecological, economic, or political system can cause a global crisis. Linear thinking and the belief that the whole is only the sum of its parts are evidently obsolete."[2] In fact, complexity has become not only an important topic but also the key to scientific explanations in all areas of science.

This does not necessarily mean that conceptual clarity has been achieved in questions of complexity. For the concept of complexity displays different (scientific) meanings depending on the area to which it is applied, even while its basic meaning remains constant. Are the concepts used in different disciplines similar, or may a phenomenon be, for instance, biologically complex but physically not?

1 K. Mainzer, *Thinking in Complexity: The Computational Dynamics of Matter, Mind and Mankind*, 5th edition, Berlin and Heidelberg: Springer 2007, p. VII.
2 *Ibid.*, p. 1.

https://doi.org/10.1515/9783110596687-001

Does the fact that some problems are in principle unsolvable for reasons of complexity (due to limited time and computational power) pose a problem for scientific practice? Shall our practice just ignore problems we cannot currently handle – or can science render apparently complex systems in simple underlying theories? Furthermore, is there a difference between the *complex* and the *complicated* such that some complex systems are not actually complicated even though all complicated systems are indeed complex. In general, again, complexity has become an important area of research in many disciplines in the last decades. For instance, the complexity and the ensuing unpredictability of weather systems has been known for a long time. And theoretical tools to master complexity have been developed in biology, where the apparent complexity of organisms has been used to argue against evolutionary theory, as well as in economics and social theory, where so-called "complexity theory" aims to help us understand systems which appear unsystematic.

As to the distinction between complexity and complicatedness[3]: The greater the number of objects and relations of a system, the greater its complexity. Complicatedness depends on the inhomogeneity of the object area. There can thus be systems of high complexity but small complicatedness (for example: organic molecules composed of numerous elements of few different kinds) whereas high complicatedness as a rule leads to complexity (for example: organisms). No wonder that the theory of complex dynamic systems, in which cause-and-effect connections are non-linear (for instance in the motion of more than two bodies under the influence of gravity), is currently becoming ever more influential, especially because of its many applications (another example the prediction of developments in the weather). This discipline closely joins newer mathematical methods such as chaos theory to older methods from statistics and probability theory.

In so far as the reduction of complexity is done in explanatory intent, this is achieved especially by model building. Models serve to simplify complex structures and to visualize abstract structures. Thus, astronomical models (for instance, in the form of orreries) were viewed in the sense of the first purpose (simplifying structures), and physical models (for instance in the form of the atomic model) were viewed in the sense of the second purpose (visualizing abstract, non-intuitive structures) and mechanical models (for instance, in the form of corpuscular models) generally in the sense of both purposes (describing visualiza-

3 The following is taken from K. Lorenz, "komplex/Komplex," in: J. Mittelstrass (ed.), *Enzyklopaedie Philosophie und Wissenschaftstheorie*, 2nd edition, vol. IV, Stuttgart and Weimar: J. B. Metzler 2010, pp. 277–279.

ble situations that were nonetheless in need of explanation by the basic concepts of space, time, mass and force). As a rule, we should differentiate between scale models, analogue models and theoretical models. Scale models are enlarged or miniature replicas of real or imaginary objects, for instance, in the three-dimensional representation of the DNA-molecule ("double helix"). Analogue models represent an object in a structurally similar (homomorphic) other object, for instance, in the form of the planetary model of the atom (in physics) or computer models of the brain in the philosophy of mind. Theoretical models consist of a set of assumptions and equations with which the essential properties of an object or system are to be grasped, for instance (in the intuitive case) in the form of Niels Bohr's atom model or of billiard balls models in the kinetic theory of gases.

As a rule a complex state of affairs cannot be completely grasped, even when models are applied. This is for instance the case where chance plays a role. The Copenhagen interpretation of quantum mechanics, that is, the theory of microphysical phenomena, assumes an irreducible, ontological contingency, that is, the existence of absolute chance in the physical world. The assumption is not uncontroversial. For instance, David Bohm's interpretation of quantum mechanics suggests that the quantum world can in fact be grasped with causal-deterministic vocabulary. From this, and from the fact that Bohm's interpretation and the Copenhagen interpretation of quantum mechanics are empirically indistinguishable[4], it follows that it may not be possible to find out whether there is really absolute chance in the world or not. All arguments for and against seem here to be relative to a physical theory and its interpretation. How are we supposed to know whether – remembering Albert Einstein's admonition that God does not play dice – there is not the possibility of a deeper deterministic description that excludes accident while coping with complexity. Not only philosophy, but natural science as well has its difficulties with chance and necessity.

Nothing is changed by the circumstance that complex relations cannot be completely grasped. This can in turn be elucidated under the concept of predictability:[5] Even in a deterministic world there are limits to predictability. Two reasons can be given in support of this. First, *deterministic chaos*. This refers to the strong dependence of a system's states on the magnitude of defined parameters. Since the magnitude of these parameters can never been known, the prediction of system's states is bound by uncertainty, which translates into a range of differ-

---

4 See J. Mittelstrass, *Konstruktion und Deutung: Ueber Wissenschaft in einer Leonardo- und Leibniz-Welt*, Berlin: Humboldt University 2001, p. 18.

5 See, in more detail, chapter 2.

ent developments in chaotic systems. Second, the problem of a *Laplace's demon.* This label (credited to Emil Du Bois-Reymond[6]) refers to a fictitious superhuman intelligence, which – under the assumption of a stable, closed and all-determined system typical for a mechanistic worldview – knows of all initial conditions of all possible movements and thus can predict the location of any particle for every point in time. Now, quantum mechanical systems are non-deterministic with regard to conjugate variables such as position and momentum. Rather, they are statistical, i. e. incalculable even by *Laplace's demon.* But whatever holds for a deterministic world also holds for a complex world and its reductions.

2. With the concept of *reduction* or reductionism philosophy of science denotes, on the one hand, an essential aspect of scientific theory formation and, on the other, a procedure that describes the successful reduction of one theory to another. In general the concept of reduction involves tracing back entities, concepts or theories to others. Reductions serve the goal of unification of the scientific world picture through the use of as uniform a conceptual system – and consequently ontology – as can be had and the elimination or replacement of philosophically or methodologically problematical concepts (or the entities they refer to) by unproblematic concepts (ontological reduction). Examples are the reduction of phenomenological thermodynamics to statistical dynamics, the reduction of Mendelian genetics to molecular genetics and the ontological reduction of psychological processes to physical processes via a theory reduction of psychology to neurophysiology.

One expression of a reductionistic programme is so-called *physicalism*, that is, the programme to express all (non-logical) expressions of a unified scientific language in the language of physics. There are two versions: "The strictest version of physicalism restricts all scientific theories to the terms of currently accepted physics. This view demands, for example, that all processes or objects can be assigned a particular quantum of energy. A weaker variant of physicalism demands the completeness of the physics of the time. This conception accordingly takes the historical change of physics explicitly into account. This view of physicalism makes a comprehensive claim for the validity of the theory of inorganic phenomena and asserts that all entities (i. e., including biological and psychological ones) are physical. A further weakening of the concept of physicalism results if only the natural sciences of the time taken as a whole are set to be comprehensive and complete. In particular, this includes the possibility that biology is not reducible to the theory of inorganic phenomena, but must have recourse to

---

6 "Ueber die Grenzen des Naturerkennens" (1872), in: E. Du Bois-Reymond, *Vortraege ueber Philosophie und Gesellschaft* (ed. S. Wollgast), Hamburg: Meiner 1974, pp. 56–57.

special regularities. In this form of physicalism (...) emergent terms and laws are admissible in principle."[7]

Now, a claim for derivability of the reduced theory from the reducing theory presupposes that both are compatible with one another. But since the reducing theory is designed to correct and improve the reduced theory, this in turn presupposes that both are incompatible. That is, the formal and informal conditions of reduction cannot be satisfied simultaneously; the correction of T_1's laws by T_2 precisely excludes their derivation.[8] This, again, is the reason why Karl Popper rejects the idea of reducibility of theories to one another and defends the incompatibility of successive theories. The principle of a critical examination characterizing a logic of scientific discovery requires, according to Popper's concept of falsifiability and the asymmetry of verification and falsification, a pluralism of theories so as to be able to select a "successful" one. Progress among theories is due to the ongoing process of critical revision of existing theories from the perspective of truth or at least verisimilitude.

3. Compared to the approaches represented in the programme of reduction, *analogies* display a weak form of relationship between entities, concepts or theories. Here the point is that this connection can be materially different but formally the same. We should distinguish between *structural* and *functional analogies:* "If the correspondence of particular relationships among the elements of a system with one another is reversibly unique to those among elements of another system (without there needing to be a correspondence between the elements themselves), we say that both systems agree partially in their structure or that a 'structural analogy' holds between them. If one grasps similarity as agreement of two systems in certain (not all) 'characters' in the sense of properties of their elements or element groups, then similar systems agree also in the relationships between the corresponding elements or element groups and are thus structurally analogous."[9] An example would be again Bohr's planetary model of the atom. A "functional analogy" between two systems on the other hand occurs if these are equally suited for a particular purpose, that is, interchangeable for achieving that purpose. An example here: the concept of force in physics and everyday

---

7 M. Carrier and J. Mittelstrass, *Mind, Brain, Behavior: The Mind-Body Problem and the Philosophy of Psychology*, Berlin and New York: Walter de Gruyter 1991, p. 172.

8 See again M. Carrier and J. Mittelstrass, *Mind, Brain, Behavior*, p. 43. Here too the proof that this difficulty has been solved by applying Tarski's concept of interpretability to the reduction problem (A. Tarski, "A General Method in Proofs of Undecidability," in: A. Tarski *et al.*, *Undecidable Theories*, Amsterdam: North-Holland 1971, pp. 1–35).

9 Chr. Thiel, "Analogie," in: J. Mittelstrass (ed.), *Enzyklopaedie Philosophie und Wissenschaftstheorie*, 2nd edition, vol. I, Stuttgart and Weimar: J. B. Metzler 2005, pp. 117–118.

life. Epistemologically speaking, both cases are forms of *similarity*, that is, agreement of two systems in some, but not necessarily all, characteristics. Analogue models accordingly represent a system or an object in a structurally similar (homomorphic) or in a functionally similar system or object.

4. The line of thought pursued here in the case of the concepts complexity, reduction and analogy lead in the philosophy of science to a position that on the one hand turns against the reductionist programme and on the other hand represents the attempt to do justice to the actual complexity of scientific objects, concepts or theories in a different manner as well, namely in the sense of a unity to be regained, a *holistic* unity of disciplinary and transdisciplinary explanations.[10] Under the *designation holism* are to be understood methodological approaches to the explanation of conceptual or empirical phenomena, that take their point of departure from a "holistic" point of view. Conceptually or methodologically, the issue is in particular the distinction between the part-whole relation and the element relation, since wholes are understood as compositions of parts but not merely as the sum of their parts. This is the case because the relations determining the composition make the whole an independent unity, whose qualities cannot be completely traced back to the qualities of the parts. The concept holism was introduced in 1926 in a biological context.[11] It also plays a role in the interpretation of quantum theory, in social-scientific theory formation and in the theory of confirmation.

In biology the concept of holism designates the attempt, in opposition to the particular positions of mechanism and vitalism, to derive all phenomena of life from a holistic "metabiological principle." According to this view biological processes can be adequately explained only if organisms are not grasped as isolated natural bodies (as in physics), but are rather seen in structure and function as standing in inseparable interaction with their own subsystems and the environment. Depending on how this abstract principle is conceptualized, it has either found general recognition in biology or been dismissed as incompatible with the biological facts. For the paleontologist Edgar Dacqué, for instance, holism was a methodological part of a teleological conception of evolution in which humankind, as the primeval form of life, included all the developmental possibilities of the animal kingdom (the animal species appear in this conception as dead ends in biological development).[12] In physics the appearance of

---

10 I follow closely here my article "Holismus" in the *Enzyklopaedie Philosophie und Wissenschaftstheorie*, 2nd edition, vol. III, Stuttgart and Weimar: J. B. Metzler 2008, pp. 427–430.

11 J. C. Smuts, *Holism and Evolution*, London: Macmillan 1926, 3rd edition 1936.

12 E. Dacqué, *Leben als Symbol: Metaphysik einer Entwicklungslehre*, Munich and Berlin: Oldenbourg 1928.

so-called entangled states in quantum theory is often viewed as a violation of the principle of separation and as the basis for an *ontological holism*. This principle states that every physical system possesses its fundamental properties independent of other systems distinct from it. The exhibition of these properties, but not their presence, can be influenced by their interactions with other systems. In composite systems the state of the aggregate system results from the states of the subsystems and their interactions.

In entangled states, such as described in the so-called *Einstein-Podolsky-Rosen paradox*, an aggregate system consisting, for instance, of two initially coupled and later separated particles has constant properties – it is in a pure state – although this does not hold for the subsystems. The aggregate system exists in a well-defined state, whereas the subsystems do not possess the correlated properties (such as spin and polarization) independently of one another. The probability distribution for the appearance of particular property values of the aggregate system cannot be calculated as the product of such probability distributions for the subsystems. Accordingly the state of the aggregate system does not supervene on the states of the sub-systems. The holism of quantum theory is expressed in the violation of the principle of separation, through which the whole is ascribed primacy before the parts.

In the philosophy of the social sciences *methodological holism* is the view that social relations can only be interpreted and explained in terms of social wholes. This holism is methodological insofar as it primarily refers to the conditions of understanding. The counter-position is so-called methodological individualism, as advocated, for instance, by Popper among others. According to this individualism all social relations can be explained out of the actions of individual persons and their interactions, which in turn can be traced back to motives and beliefs and thus need not necessarily refer to social wholes. Opposed to this position, advocates of holism such as Karl Marx and Émile Durkheim postulate the impossibility of abstracting from the influence of social institutions on the behaviour of individuals. According to Marx social conditions and their development can only be interpreted in categories of social "totalities" such as relations of production or classes; for Durkheim institutions such as family or religious communities act as social facts upon the individual.

While biological, quantum-physical and social-scientific elaborations of holistic notions are supposed to serve the particular interpretative and explanatory needs of partial areas of investigation, the so-called *confirmation holism* of philosophy of science deals with the over-arching thesis that theories can only be empirically evaluated as wholes. This form of holism arises in the framework of a hypothetical-deductive conception of empirical testing and confirmation, according to which the investigation of the tenability of a hypothesis, not comprehensively testable by immediate observation, is carried out by deriving empirically accessible consequences. If the consequences turn out to be true, the hypothesis is taken to be empirically confirmed. However, Pierre Duhem pointed out that the derivation of empirical consequences must have recourse to numerous other hypotheses, for instance, those taken from background knowledge or those about the function of the measuring instruments applied. Every successful test confirms not only the hypothesis under consideration, but also the entire group of hypotheses used in the testing process.

Similar arguments are made in philosophy of science in the framework of so-called *meaning holism* or *semantic holism*. Here, the meaning of individual concepts or propositions results from their interactions with other linguistic structures. They do not have meaning in isolation but only in the context of comprehensive language systems. This holism arises out of confirmation holism when it is joined to the verificationist premise that the conditions of empirical testing provide information on meaning. A further ground is the realization that the meaning of scientific concepts is understood only in the context of the corresponding scientific theory and cannot be acquired by knowledge of the appropriate definitions alone. The meaning of a concept like that of force can be clarified only by the role that it plays in the system of the laws of mechanics. According to this context theory of meaning a scientific concept acquires its specific content only through its integration in theory.

As an aside let me remark that holistic approaches of this kind lead to the concept of *emergence* insofar as, both in the sense of the confirmation holism and also in the sense of semantic holism, it is the system-properties that give us information about the behaviour of the system. These properties are in turn *emergent.*[13] Emergence says again that it is impossible to use characteristics of elements and the interrelations between these to describe characteristics of ensembles or make predictions about them. The core element of a *strong* emergence thesis is a non-derivability or non-explainability hypothesis of the system characteristics shaped from the characteristics of the system components. An emergent characteristic is non-derivable; its occurrence is in this sense unexpected and unpredictable. *Weak* emergence is limited to the difference of the characteristics of systems and system components and is compatible with the theoretical explainability of the system characteristics. Weak emergence in turn is essentially a phenomenon of complexity.

Here, too, our considerations return us to the concept of complexity, which is, from the perspective of philosophy of science as well, the key concept of the

---

13 See, in more detail, chapter 2.

modern development of science and points to the future, possibly also to the limits, of scientific progress.

2 Predictability, Determinism, and Emergence

Humans are creatures for whom the future is part of present existence, bounded by uncertainty in many respects, but indispensable for comprehending the present. Immanuel Kant views the "anticipation of the future" as the "most decisive proof of man's advantage, in that he is able to prepare for remote objectives in keeping with his destiny."[1] And for Martin Heidegger, the structure of human existence is future oriented in itself.[2] In one sense this holds for ordinary experience, as reflected in anthropological studies, and in yet another sense it holds for science and leads – in connection with the original Greek idea of order in the physical world – to epistemological analysis. In both areas, predictability is the attempt to deal with the future, and in science – for example in the thesis of the structural identity of explanation and prediction – it is also a crucial criterion of a theory. Predictions serve as both an application of a theory and as its confirmation. The following discussion is limited to addressing the problems connecting to these scientific issues.

1. Problems with predictability in science have been discussed for a very long time. This is particularly so for complex relationships. A classic example is the hole in the ozone layer, or, the effects of chlorofluorocarbons (CFCs) on the high atmosphere ozone layer. In this case, the causal relationships of the chemical reactions are so complex that it is almost impossible to predict their effects. After all, it was difficult enough to explain the mere occurrence of the effect. Just as well, it is a common fact that small causes can have large, unpredictable effects. Ice ages, for example, according to recent scientific research, are caused by a relatively minor cooling down in the earth's atmosphere. This, in turn, is caused by a decreased intensity in the rays of the sun, which results from peculiarities of the earth's revolving around the sun, in particular its varying eccentricity as well as variations in its orientation and the gradient of the earth's axis. The crucial point is that this trifling cooling down leads to a change in flow in the North Atlantic. In particular, the warm flow, which comes to the surface near Iceland and is responsible for the warm climate in Europe, is diverted. This leads to a much harsher climate in the north, which, in turn, contributes

---

1 I. Kant, "Mutmasslicher Anfang der Menschengeschichte" (1786), in: *Kant's gesammelte Schriften*, vol. VIII, Berlin and Leipzig: Walter de Gruyter 1923, p. 113 ("Conjectures on the Beginning of Human History," in: H. Reiss [ed.], *Kant: Political Writings*, 2nd edition, Cambridge etc.: Cambridge University Press 1991, p. 225).

2 M. Heidegger, *Sein und Zeit* (1927), 14th edition, Tuebingen: Max Niemeyer 1977, §§ 67 ff. ("Zeitlichkeit und Alltaeglichkeit").

https://doi.org/10.1515/9783110596687-002

toward cooling at the global dimension.[3] Thus small changes in the conditions cause, in this case, considerable changes in the state of the system as a whole.

Another example is related to Max Planck's (epistemologically problematic) exploration of free will, which has recently become relevant again for brain science. Embarking from the concept of causal universality, i.e. the assumption of causal closure of the world, Planck argues that the will is also causally determined, although mental events, e.g. thoughts, are unpredictable – even for an ideal observer – due to their manifold dependencies. For Planck, this is also relevant for the relations between a willing and a perceiving self (the ideal observer): "Each new observation (...) gives rise to a new motive, and the recognition of this motive in turn creates a new situation. The series is infinite, and since the observed person (the willing ego) owes no obedience to the observer (the percipient ego), we shall never be able to claim with certainty that the eventual decision must be in the sense of the observer's latest discovery."[4] This has, following Planck, no bearing on the continued validity of a causal law.

2. On this topic, the most commonly discussed example is chance in quantum mechanics. Quantum mechanics imposes serious limitations on the predictability of events. The central principle of the theory is *Schroedinger's equation*, which serves to determine the "state function" or "wave function" of a quantum system. The state function is generally taken to provide a complete description of quantum systems; no properties can be attributed to such a system beyond the ones expressed in terms of the state function. Erwin Schroedinger's equation determines the time development of the state function unambiguously. In this sense, quantum mechanics is a *deterministic* theory.

However, apparently irreducible chance elements enter when it comes to predicting the values of observable quantities. The measurement process in quantum mechanics is described as the coupling of the quantum system to a particular measuring apparatus. Schroedinger's equation yields, then, a range of possible measuring values of the quantity in question, each of these values being labelled with a probability estimate. That is, Schroedinger's equation only provides a probability distribution and does not anticipate particular observable events. Quantum mechanics is extended to actual measuring values by adding the so-called "projection postulate." This postulate is independent of Schroedinger's equation and says that one of the possible measuring values

---

3 See W. S. Broecker and G. H. Denton, "Ursachen der Vereisungszyklen," *Spektrum der Wissenschaft* 3 (1990), pp. 88–98.

4 M. Planck, *Vom Wesen der Willensfreiheit*, 2nd edition, Leipzig: J. A. Barth 1937, p. 18 (*The Universe in the Light of Modern Physics*, 2nd edition [with a section on Free-Will], London: George Allen & Unwin 1937, p. 101).

is assumed in actuality. The spectrum of possible values collapses into the one value that is obtained in the measurement. In repeated measurements of the same kind, the relative frequencies of the values coincide with the probability estimates supplied by Schroedinger's equation.

The salient point is that, according to present lights, this collapse of the state function, i.e., the selection of the actual measuring value from the range of possibilities is a genuinely *indeterministic* process whose outcome cannot be predicted on any basis whatsoever. These obstacles to prediction, as they become manifest in quantum mechanics, have nothing to do with the ignorance of the prevailing initial conditions. Given a complete description of the quantum state, chance fluctuations at the level of observables will yet occur. Quantum mechanics involves in-principle limitations of predictability to the effect that, for instance, it is objectively indeterminate when a given radioactive nucleus will decay. Such limitations are not merely epistemic constraints, but rather represent an ontological indeterminateness.

Heisenberg's so-called indeterminacy relations are a consequence of Schroedinger's equation, although historically they were formulated independent of this equation and prior to its enunciation. The Heisenberg relations place severe limitations on the simultaneous measurement of what is called "incompatible" or "incommensurable" quantities like position or momentum or spin values in different directions. The more precise one of the quantities is evaluated, the more room is left for the other one. Like the constraints mentioned before, the limitations set by the Heisenberg relations have nothing to do with practical impediments to increasing measurement accuracy that might overcome by improved techniques. Rather, the relations express limitations set by the laws of nature themselves.

Heisenberg's indeterminacy relations entail serious restrictions of the prediction of future quantum states. For ease of illustration consider the following spin measurements. Spin states are quantized; they possess only two possible values in each direction, namely, "spin up" or "spin down." A beam of electrons can be "spin-polarized" by sending the particles through a suitably shaped magnetic field (a Stern-Gerlach apparatus). That is, the spin of all electrons in, say, *x*-direction after exiting from the setup is, say, "up." This result can be confirmed by a second measurement of the same quantity performed directly after the first. 100% of the electrons come out "spin up" in the *x*-direction. Let the beam then pass through the same setup but now measuring the spin values in the *y*-direction, perpendicular to *x*. The outcome is that one half of the beam exhibits "spin up" and the other half "spin down." If the beam is finally sent through the apparatus this time oriented again in *x*-direction, the perplexing result is that 50% of the electrons are registered "spin up" and "spin down," respectively.

Correspondingly, the first measurement, in spite of its quite unambiguous result, could not be utilized for a prediction once a measurement of an incompatible quantity has been carried out. Again, this is a matter of principle. There is no way of anticipating the joint values of incompatible quantities below the threshold set by the Heisenberg relations. As a result, inherent limitations prevent us from predicting the future states of such quantities.

This element of genuine, irreducible chance troubled Albert Einstein very much. Einstein accepted statistical accounts if they could be viewed as growing out of incomplete knowledge of the relevant conditions and states. Quantum mechanics differed from all other statistical theories in physics in that the invocation of probability could not be attributed to human ignorance. Einstein's commitment to a determinist world was his chief reason for dissenting from quantum mechanics. As he wrote to Max Born, he found the idea "unbearable" that an electron decides on its own in which direction to move. If this turned out to be true he preferred to be an employee in a gambling casino rather than a physicist.[5] In the same vein, Einstein told Born that quantum mechanics does not bring us closer to God's mystery. After all, God does not throw dice.[6] This episode bears witness to the fact that in-principle constraints on predictability represent a serious deviation from the notion of *Laplace's demon* which is the core element of the traditional, ignorance-focused account of chance and probability.

To repeat once more: Current wisdom holds there are fundamental processes in the quantum world that inhibit randomness, which implies general limits of predictability. Nevertheless this is by and large irrelevant to macroscopic phenomena; with large numbers of atoms the uncertainties average themselves out. This, in turn, brings us to the fundamental question of the relationship between *determinism* and *predictability.*

3. Talking about the limits of predictability in principle immediately poses questions for a *deterministic world.* This has been clear to Max Planck, leading to the insight that the classical dictum "an event is causally conditioned, if it can be predicted with certainty"[7] cannot be maintained, moreover, one is forced "to acknowledge the following sentence as a given fact: In no circumstance is it possible to predict a physical event with exactness."[8] In a similar vein, several

---

5 Einstein to Born (April 29, 1924), in: *Albert Einstein – Hedwig und Max Born: Briefwechsel 1916–1955*, Munich: Nymphenburger Verlagshandlung 1969, p. 118.

6 Einstein to Born (December 4, 1926), *ibid.*, p. 129.

7 M. Planck, "Die Kausalitaet in der Natur" (1932), in: M. Planck, *Vortraege und Erinnerungen*, 5th edition, Stuttgart: Hirzel 1949, p. 252.

8 *Ibid.*, p. 253.

years before (1927), Werner Heisenberg claimed that, "in principle," quantum mechanics has the effect that, "the law of causality is in a sense unfounded. Since one can never know precisely the initial conditions, one can never calculate the mechanical course of events. (...) Concerning the sharp version of the law of causality: If we know the present, we can calculate the future – it is not the consequent, but the antecedent that is wrong."[9] This, however, is not the last word on the possibility of a deterministic world. It is rather necessary to separate the concepts of determinism and predictability from each other; determinism understood here (following John Earman) as the thesis that, if two possible worlds are identical at a given point in time, then they are identical at every point in time.[10] This does not exclude hindrances to predictability for a given state of affairs and deterministic development. The thesis is: *Even in a deterministic world there are limits of predictability.*

Two reasons can be given in support of this. First, *deterministic chaos*. This refers to the strong dependence of a system's states of affairs on the magnitude of defined parameters. Since the magnitude of these parameters can never be known, the prediction of system's states of affairs is bound by uncertainty, which translates into a range of different developments in chaotic systems. Unpredictability as a result of chaos is not limited to complex systems, rather, it can also occur in simple systems that only consist of a few elements. For example, two coupled pendulums constitute a simple system, the relevant laws of which have been known for centuries. But it has only recently become clear that, within such an arrangement, in a distinct range of initial conditions – namely system stimulations of medium strength – there can be chaotic and unpredictable oscillations. Another example, already introduced in the beginning, is meteorology, which was the original impulse for studying chaotic effects in dissipative systems. In a well-known metaphor: Even the flapping of a butterfly's wings can crucially effect the convection currents in the earth's atmosphere and, hence, meteorological developments ("butterfly effect").[11] The reliability of weather forecasts is not only constrained by *practical* limits but also by limits *in principle*. These occur even though the underlying laws are known and of a deterministic nature.

---

9 W. Heisenberg, "Ueber die Grundprinzipien der Quantenmechanik," *Forschungen und Fortschritte* 83 (1927), p. 83 (= W. Heisenberg, *Gesammelte Werke / Collected Works*, vol. I, Munich and Zurich: Piper 1984, p. 21).

10 See J. Earman, *A Primer on Determinism*, Dordrecht etc.: Reidel 1986, p. 14 ("if two worlds agree for all times on the values of the conditioning magnitudes and if they agree at any instant on the values of the other magnitudes, then they agree at any other instant").

11 See H. G. Schuster, *Deterministic Chaos: An Introduction*, Weinheim: Physik-Verlag 1984, p. 2.

More generally and again using the example of weather forecasting, this can be formulated as follows[12]: It is possible to know the exact equations of motion for a system, without being able to predict the evolution in time of this system. Although meteorological developments (as it is generally understood) can be completely described by thermodynamic equations, this is of little help. Because all observations are always finitely accurate, the future behaviour cannot be predicted using these equations. Though weather can be predicted in the short run; the chaotic effects described here will still appear in the middle to long run. It is important to see that it is not the system itself that behaves chaotically; its development is, quite the contrary, strictly deterministic. A chaos exists only for us, not for the thing being studied; it results from the imprecision of our knowledge of the initial conditions. But this means that there is an *epistemological limit* that occurs in the phenomenon of deterministic chaos. Although a system is, in fact, strictly deterministic and can be completely understood according to certain laws, we are not in a position to describe the behaviour of this system, despite our precise knowledge of these laws.

Epistemologically speaking, the chaos is a *supervenient* characteristic.[13] A characteristic *s* is supervenient to a set of physical characteristics *p*, if (1) *s* is not of a concrete-physical nature, that is, *s* can obtain in physically different systems, and if (2) differences in *s* always coincide with differences in *p* (although not *vice versa*). The occurrence or non-occurrence of chaos always depends on the physical differences in the system.

The second reason is the problem of a *Laplace's demon*. This label (credited to E. H. Du Bois-Reymond[14]) refers to a fictitious superhuman intelligence, which – under the assumption of a stable, closed and all-determined system typical for a mechanistic worldview – knows of all initial conditions of all possible movements and thus can predict the location of any particle for every point in time. Now (as has already been mentioned), quantum mechanical systems – in contrast to relativistic physics, where differential equations describe deterministic

---

12 This is also the conclusion of a longer argument in: M. Carrier and J. Mittelstrass, *Mind, Brain, Behavior: The Mind-Body Problem and the Philosophy of Psychology*, Berlin and New York: Walter de Gruyter 1991, p. 262. Cf. M. Carrier, "Chaostheorie," in: J. Mittelstrass (ed.), *Enzyklopaedie Philosophie und Wissenschaftstheorie*, 2nd edition, vol. II, Stuttgart and Weimar: J. B. Metzler 2005, pp. 40 – 43.

13 See P. Hoyningen-Huene, "supervenient / Supervenienz," in: J. Mittelstrass (ed.), *Enzyklopaedie Philosophie und Wissenschaftstheorie*, vol. IV, Stuttgart and Weimar: J. B. Metzler 1996, pp. 144 – 145.

14 "Ueber die Grenzen des Naturerkennens" (1872), in: E. Du Bois-Reymond, *Vortraege ueber Philosophie und Gesellschaft* (ed. S. Wollgast), Hamburg: Meiner 1974, pp. 56 – 57.

systems with regards to their state variables – are non-deterministic with regard to conjugate variables such as position and momentum. Rather, they are statistical, i.e. incalculable even by *Laplace's demon* – an implication confirmed by recent developments in physics.

There is yet another reason why *Laplace's demon* is unable to handle the problem of predictability, even under the assumption of deterministic structures.[15] Such a demon would himself be part of the world which he seeks to predict. This situation inhibits *self-reference:* the observing system or measurement device is itself part of the system whose development is being predicted. In other words, predictability in a *Laplace's demon* situation demands measurability of a system state "from within." This, in turn, demands, (1) that any object state is connected to the state of an apparatus, hence there can be no object states remaining (though it is possible that apparatus states can exist without a corresponding object state), and (2) that there are no two object states that correspond with the same apparatus state (while conversely there can be two apparatus states which correspond to the same object state).[16] To measure each object state with exactness, it must correspond to at least one apparatus state. This implies first and foremost that there are at least as many apparatus states as there are object states. However, the assumption of the *inner observer* (a demonic situation) implies that there are more object states than apparatus states. An inner observer is indeed part of an object, such that for the inner observer the apparatus states are a proper subset of the object states.

These conditions contradict each other. The demand for exact measurement implies that the apparatus has at least as many states as the object. The condition of the inner observer says that the object has more states than the apparatus. These two conditions cannot hold at the same time. And this is a strong argument for the separation of predictability and determinism. Both arguments, the chaos argument and the argument of the inner observer, make it clear that there can be deep or even basic problems of prediction even in a deterministic framework; hence determinism and unpredictability are not mutually exclusive.

4. *Laplace's demon* has lost its demonic character in this context; he has become an observing scientist. Thus it has been rightly said: "In fact, most of the contributors to the debate, having paid lip service to Laplace, almost unnotice-

15 For the following see Th. Breuer, *Quantenmechanik: Ein Fall fuer Goedel?*, Heidelberg etc.: Spektrum Akademischer Verlag 1997, pp. 7–21; "Limits to Self-Observation," in: M. Carrier *et al.* (eds.), *Science at Century's End: Philosophical Questions on the Progress and Limits of Science*, Pittsburgh Pa.: University of Pittsburgh Press 2000, pp. 135–149.

16 In Th. Breuer's presentation this is expressed by the subjectivity of the picture ("Can a picture contain a full and precise picture of itself?"), "Limits to Self-Observation," p. 135.

ably substitute for his demon a human observer. They thereby reduce determinism to predictability, i.e., the question whether an actual observer, a biologist or a physicist, is able to predict future events. This reduction of Laplacian determinism to actual predictability is a drastic step. On the one hand, it brings the question from philosophical clouds down to earth, where one may hope to find an answer. On the other, it is reduced to a technical question about the state of affairs in the relevant science."[17] This is also the case with Karl R. Popper.

For Popper, who by and large identified determinism and predictability with each other (determinism = predictability with a defined level of exactness, which depends on the degree of knowledge about the initial conditions)[18], *scientific determinism* is "the doctrine that the state of any closed physical system at any given future instant of time can be predicted, even from within the system, with any specified degree of precision, by deducing the prediction from theories, in conjunction with initial conditions whose required degree of precision can always be calculated (in accordance with the principle of accountability) if the prediction task is given."[19] In this context, Laplace's demon is construed as a disembodied spirit; he is transformed into a "super-scientist:" "The demon, like a human scientist, must *not* be assumed to be able *to ascertain initial conditions with absolute mathematical precision*; like a human scientist, he will have to be content with a finite degree of precision."[20] Naturally, this leaves room also for deterministic conceptions.

Popper's critique of determinism in natural science and philosophy employs arguments not only from quantum mechanics, but also from classical physics. He argues that Newtonian mechanics, which is deterministic by conception, is unable to determine initial conditions with the precision necessary for prediction ("principle of accountability"). More generally, according to Popper, the growth of theoretical knowledge is not predictable in principle, which also hints at an *indeterminism* – which could be used, for example, as a solution for mind-body problems.

5. Finally, the concept of *emergence*. Emergence says that it is impossible to use characteristics of elements and the interrelations between these to describe

---

17 N. G. van Kampen, "Determinism and Predictability," *Synthese* 89 (1991), p. 275.

18 See K. R. Popper, "Indeterminism in Quantum Physics and in Classical Physics," *The British Journal for the Philosophy of Science* 1 (1951), pp. 117–133.

19 K. R. Popper, *The Open Universe: An Argument for Indeterminism*, Totowa: Rowman & Littlefield 1982, p. 36.

20 *Ibid.*, p. 34. Cf. J. Earman, *A Primer on Determinism*, pp. 8–10.

characteristics of ensembles or make predictions about them.[21] Thus a common formula says this: the whole is more than its parts.[22] According to the *emergence thesis*, the world is a levelled structure of hierarchical organized systems, where the characteristics of higher-level systems are by and large fixed by the characteristics of their respective subsystems, yet at the same time essentially different. Different characteristics and processes occur in the respective levels. As well, a weak and a strong emergence thesis can be distinguished from one another.

The core element of the strong emergence thesis is a non-derivability or non-explainability hypothesis of the system characteristics shaped from the characteristics of the system components. An emergent characteristic is non-derivable; its occurrence is in this sense unexpected and unpredictable. Weak emergence is limited to the difference of the characteristics of systems and system components and is compatible with the theoretical explainability of the system characteristics. Weak emergence is essentially a phenomenon of complexity.

The classic rendering of strong emergence is credited to Charlie D. Broad.[23] Broad's motivation was to provide a suitable interpretation of living organism. He intended to depict organisms neither as mere machines nor as being fuelled by an exceptional vital force. This neo-vitalist view was first and foremost endorsed by Hans Driesch[24], who maintained that beings are fitted with "entelechy," i.e. with purposeful biological powers. Broad was searching for a third way between the mechanistic and the vitalist view on life. The emergence thesis was intended to create this path. Emergent characteristics of ensembles were intended to be roughly defined by the divergent characteristics of their components, yet it was not intended to explain the former on that basis.

Strong emergence is characterized through the following conditions: (1) The condition of *qualitative difference*. This condition applies the emergence thesis to those characteristics of ensembles which differ profoundly from the characteristics of their components. (2) The condition of *characteristic determination*. This condition says that the characteristics of the components are sufficient to let the specific characteristic emerge; emergence is not depended upon further factors. (3) The condition of the *principal gap of explanation*. This condition implies

---

21 For the followings see M. Carrier, "emergent / Emergenz," in: J. Mittelstrass (ed.), *Enzyklopaedie Philosophie und Wissenschaftstheorie*, 2nd edition, vol. II, Stuttgart and Weimar: J. B. Metzler 2005, pp. 313–314.

22 See K. Lorenz, "Teil und Ganzes," in: J. Mittelstrass (ed.), *Enzyklopaedie Philosophie und Wissenschaftstheorie*, vol. IV, Stuttgart and Weimar: J. B. Metzler 1996, pp. 225–228.

23 Ch. D. Broad, *The Mind and Its Place in Nature*, London: Routledge & Kegan Paul 1925.

24 H. Driesch, *The Science and Philosophy of Organism*, vols. I-II, London: Adam and Charles Black 1908, 2nd edition, London: Adam and Charles Black 1929.

that it is actually impossible to explain the characteristics of ensembles through the characteristics of their components, including their interrelations. – Incidentally, the existence of strong emergent characteristics in this sense is heavily disputed. The only candidates in the running at the moment are currently *phenomenal* characteristics.[25] The point here would be, that in a given neurophysiological arrangement, the occurrence of qualitative experiences (e.g. blue, the sound of trumpets etc.) in a system would be non-derivable and unpredictable.

Concerning predictability, it is particularly the *temporal* aspect of the emergence thesis which is of interest, i.e. for ensemble characteristics that occur in developments. Limits of reductability (of the whole to its parts) figure here as limits of explanation and predictability. This temporal novelty is described by the concept of *creative advance of nature*. It is endorsed by Popper and Eccles, among others.[26]

6. To sum up: Determinism does not imply predictability, and unpredictability does not imply non-determinism. In fact, there is unpredictability in a deterministic world, and unpredictability permits deterministic worlds. This has been illustrated with the discussion of the concepts of deterministic chaos, *Laplace's demon*, who becomes stripped of all his demonic characteristics, and emergence. Besides, one could not simplify matters by distinguishing (as has been proposed) between ontological determinism and epistemic unpredictability. First, such a distinction is epistemic in itself and second, it merely expresses that the concepts of determinism and predictability do not belong to the same (semantic) level, or even mean the same thing. Predictability, not determinism, is the problem (in some areas). Dealing with unpredictability in the right way is the challenge – in science as well as in ordinary life.

---

25 See A. Stephan, "Phaenomenale Eigenschaften, phaenomenale Begriffe und die Grenzen Reduktiver Erklaerung," in: W. Hogrebe and J. Bromand (eds.), *Grenzen und Grenzueberschreitungen. XIX. Deutscher Kongress fuer Philosophie, Bonn, 23.–27. September 2002. Vortraege und Kolloquien*, Berlin: Akademie Verlag 2004, pp. 404–416.

26 K. R. Popper and J. C. Eccles, *The Self and Its Brain*, New York etc.: Springer International 1977, pp. 22–35. Cf. also M. Čapek, *The Philosophical Impact of Contemporary Physics*, Princeton N.J. etc.: Van Nostrand 1961, pp. 333 ff.

3 Discovery

What is known to science is known in different ways – a quick look at, for instance, textbooks in mathematics, physics, biology or economics renders this apparent. The same is true of the ways in which science acquires its knowledge. There are inductive, deductive, experimental and many other methods. And its successes are documented in confirmed hypotheses, explanations of the hitherto inexplicable, as well as in discoveries.

It is above all the discoveries which represent the appeal of science – for the scientific layman as well as for the scientist. The new is revealed in discoveries, and the new is, by all means, the aim of scientific endeavours. Aiming at the discovery of the new, science develops its methods and structures and defines its concept of research. Thus, occasionally a distinction is made between hypothesis-driven and discovery-driven research, apparently thinking of the expected new in the former, and of the unexpected new in the latter case. But such a distinction is artificial. After all, hypothesis-driven research is also aiming at discovering the new, and discovery-driven research also requires methodic tools, parts of which are in turn hypotheses. Precisely this is what the history of science teaches, provided one conceives of it not just as an arsenal of past insights and errors, but also as an expression of the scientific spirit, which recognizes itself in past as well as present representations.

Even science does not have a subscription to the new and no strategy for finding it routinely. There are many ways leading to Rome, but just as well to new scientific insights – and, of course, many leading past them too. To put it in some more detail, there are three paths on which new insights have been found in the history of science, using short examples: (1) Discoveries which were surprising to science and the discoverers themselves, (2) discoveries which had been expected by the discoverer and science up to a point, but which were novel in their details, (3) discoveries which had been expected by the discoverer, but which came as complete surprise to science.[1]

First, the discoveries which came as surprises to science and even to the discoverer himself, hence representing something unexpected new. Two examples: (1) The first is very famous: the *discovery of the X-ray.* In November 1895, Wilhelm C. Roentgen was experimenting with a gas discharge tube, that is, an instrument that emits electrons (for instance, from a heated wire), and accelerates them

---

1 A comprehensive account of the development of the new in science may be found in: J. Mittelstrass, *Leonardo-Welt: Ueber Wissenschaft, Forschung und Verantwortung,* Frankfurt: Suhrkamp 1992, pp. 74–95 (I.4 Die Wissenschaften und das Neue).

https://doi.org/10.1515/9783110596687-003

through applying voltage; the tube itself is under vacuum, or at least, the gas pressure is significantly reduced. He covered the tube with cardboard paper to see to what extent it would still let light pass, and thus observed that some crystals, lying on his desk for no particular reason, started to fluoresce. Roentgen discovered that this fluorescence must have been caused by a new sort of radiation, being emitted by the gas discharge tube, and apparently able to cover great distances. This strikingly powerful radiation were X-rays, today well-known to us.

Second, Rutherford's *discovery of the atomic nucleus.* In 1909 Ernest Rutherford was doing an experiment to examine the structure of the atom, which was designed following military rules: If you don't know what the object in front of you is, then you'd better shoot at it.[2] Rutherford used a radioactive material as ray gun, and shot a narrow ray of alpha-particles on a thin metal sheet. Behind this sheet, a fluorescent screen had been mounted, which, when an alpha-particle would hit it, would document this with a microscopic flash of light. The big surprise, now, was that the alpha-particles were not just observed to be slightly redirected after hitting the sheet, but that their direction was changed altogether. Some even were repelled by the sheet, as if they had hit a solid wall. Rutherford later observed: "It was quite the most incredible event that has ever happened to me in my life. It was almost as incredible as if you fired a 15-inch shell at a piece of tissue paper and it came back and hit you."[3] Rutherford then tried to do justice to this "incredible event" by developing his theory of the structure of the atomic nucleus. Some of the (positively charged) alpha-particles had directly hit the (also positively charged) atomic nuclei and were hence repelled by them in their initial direction of movement.

Such are the examples for the entirely unexpected discoveries, for the entirely unexpected new. As examples for the discoveries which had been expected, to some extent, by the discoverer as well as by science in general, but which were novel in detail, let us consider the following two: First, Lavoisier's *discovery of the composite nature of water.* Antoine Laurent de Lavoisier had, since 1778, been looking for the oxide (the compound with oxygen) of hydrogen, discovered in 1766 by Henry Cavendish, but had not made any tangible progress. It was only in 1781, when Cavendish noticed (and Lavoisier learned of this in 1783), that in an explosion of so-called detonating gas, hydrogen and oxygen could be permuted into their own weight in water, that Lavoisier inferred that the long searched-for

---

2 See R. U. Sexl, *Was die Welt zusammenhaelt: Physik auf der Suche nach dem Bauplan der Natur,* Frankfurt and Berlin and Vienna: Ullstein 1984, p. 145.

3 E. Rutherford, "The Theory of Atomic Structure," in: J. Needham and W. Pagel (eds.), *Background to Modern Science: Ten Lectures at Cambridge Arranged by the History of Science Committee 1936*, Cambridge: Cambridge University Press 1938, p. 68.

hydrogen-oxide was really water itself, and that hence water was not an elementary substance, but a compound. So, what was not surprising here was the existence of a hydrogen-oxide; entirely surprising was the discovery that this oxide was the well-known substance water.

Second, Oerstedt's *discovery of electromagnetism*. In the natural philosophy of romanticism, all natural powers were thought of as expressions of a single and fundamental force, and this gave rise to the expectation that natural powers would have to be convertible into each other. In particular, this was thought to apply to electricity and magnetism. Under the influence of this idea, Hans Christian Oerstedt searched for such a conversion and discovered, in 1820, more or less accidentally on the occasion of a lecture, that a magnetic needle would be deflected by a wire conducting electricity. Thus, the connexion between electricity and magnetism was discovered. The novel and unexpected aspect of this effect, now, consisted of the fact that the current would cause the magnetic needle to rotate. What had been looked for was a direct attraction or repulsion between the electric charge and the magnetic poles. What had not been expected was a circular magnetic field surrounding the conductor the current was flowing through. A mistaken background assumption had prevented the discovery of electromagnetism for a while; chance had to come to the rescue to lead to the right track.[4]

Finally, two examples for discoveries which had been expected by the discoverer, but came as complete surprises to science. First, *Poisson's white spot*. In about 1830 Augustin J. Fresnel presented the first complete version of a wave theory of light. The heart of this theory was the assumption that waves of light are of a transversal nature (that is, they occur perpendicularly to the direction of propagation). Siméon D. Poisson thought this theory to be wholly absurd, and to prove this, he deduced an apparently nonsensical consequence from it. According to Fresnel's theory, a white spot would have to occur at the midpoint of the shadow produced by a point-shaped source of light directed at a circular disc (*Poisson's white spot*). François Arago then undertook the (to all appearances, redundant) labour to also demonstrate the falsity of this consequence experimentally. But, entirely surprisingly, he really did find the predicted spot. Although it was not the originator of the theory who predicted this novel effect, he could have done so, had he executed the deduction. In cases of a the-

4 See R. U. Sexl, *Was die Welt zusammenhaelt*, pp. 61–62. For a diverging account see J. Agassi, *Towards an Historiography of Science*, The Hague: Mouton 1963, pp. 67–74. Agassi assumes, that the discovery was less accidental than traditionally thought.

oretical prediction of a novel effect, its empirical manifestation might surprise the general public, but not the theorist.

Second, Einstein's *prediction of the deflection of light in the gravitational field.* Einstein had deduced, from the *general theory of relativity*, that light would be deflected from its initial trajectory by a certain angle by a gravitational field. In 1919, Arthur S. Eddington examined the position of stars, the light of which was passing the sun very closely. He then compared these positions with those of the same stars at times these were more distant from the sun. Indeed, the predicted deflection was observed in the precise degree expected. This successful theoretical prediction was greeted with great surprise by the uninitiated. At the same time, the process of testing the *general theory of relativity* has been compared to the Catholic procedure of canonization ("November 6, 1919, the day on which Einstein was canonized"[5]). The successful prediction of the deflection of light was one of the required miracles.

These examples teach that there is no simple way to the new in science, and that the diverse ways to the new are not simple. Furthermore, they are only rarely due to the strict following of scientific programmes, and this is why talk of scientific revolutions, which has again become popular with Thomas S. Kuhn's work in the history of science, is not that misguided. Indeed, scientific revolutions differ from political and social ones by having a much more varied potential for change, and, as a rule, for producing fewer losses, but, after all, this is not a disadvantage and not deplorable. Moreover, theories in science sometimes die more honourably than on the political or ideological stage. There, they often only come to an end with the biological end of their adherents – although this is not altogether unknown in science too.[6]

Science, in its search for the scientific new, does not just get driven by discoveries, accidental or non-accidental ones – one could express this as striding from truth to truth – but, surprisingly, also via errors, in a heuristic sense.[7] Let us take a final example also for this heuristic fruitfulness of errors, Einstein's *derivation of the general theory of relativity.* This relied on principles which were par-

5 A. Pais, *'Subtle is the Lord ...': The Science and the Life of Albert Einstein*, Oxford: Clarendon Press and New York: Oxford University Press 1982, p. 305.

6 See M. Planck, *Wissenschaftliche Selbstbiographie*, Leipzig: J. A. Barth 1948, p. 22 ("A new scientific truth does not usually become accepted by having its opponents convinced and having them declare their new conviction, but mostly by its opponents dying out, and having the new generation getting acquainted with truth straightaway" [my translation]).

7 Compare the more comprehensive account in: J. Mittelstrass, *Die Haeuser des Wissens: Wissenschaftstheoretische Studien*, Frankfurt: Suhrkamp 1998, pp. 13–28 (I.1 Vom Nutzen des Irrtums in der Wissenschaft).

tially motivated epistemologically. One of the principles pertinent and even essential for the formulation of the theory is *Mach's Principle*, as Einstein calls it. According to the ideas of Isaac Newton, there is one prime, indeed, truly immobile, system of reference, "absolute space;" movements relative to absolute space are indicated by the presence of forces of inertia (for instance, centrifugal forces). Through such forces, accordingly, absolute space takes effect on the objects, while, at the same time, the objects are never able to disturb absolute space in its tranquil existence. Einstein considered the assumption of a uni-directional causation inconsistent and assumed instead that the forces of inertia are explicable through the relative positions and movements of the bodies (an idea he attributed to Ernst Mach).

Mach's principle is not just one of the central motives for the development of the *general theory of relativity*, but also plays an important role in the process of formulating the theory. However, it turns out that the fully developed theory does not satisfy *Mach's principle*. That is, the theory allows space-time structures as physically possible in which forces of inertia originate out of an overarching space-time that is independent of objects and their movements – even if this does not have the Newtonian shape of an absolutely immovable space. Furthermore, according to our current empirical knowledge, we may assume that in our universe one such space-time structure contradicting *Mach's principle* has been realized. Hence, *Mach's principle* would be false; following today's theoretical and empirical state of knowledge, our universe is equipped with a space-time structure that is in part independent of the mass-energy distribution in the universe. Nevertheless, *Mach's principle* played an essential and probable indispensable role in the process of formulating the *general theory of relativity*. At least, the development of the theory would not have been possible on the path taken by Einstein without assuming this principle.

In other words, error too may play an essential role, not just not hindering scientific progress, but even furthering it. The game of science (Karl R. Popper) knows many successful moves; one of these is the truth of error.

4 Time

Time is an unending theme of philosophical reflection. This is because time is ever present and, simultaneously, something puzzling: time in our experience is never constantly the same – sometimes it passes quickly, sometimes slowly, sometimes it passes by, sometimes it changes us – because time, especially in connection to space, leads us to the foundations of natural science and to its beginnings. Because time holds open the door between myth and reality, questions about beginnings in time easily become questions about the beginning of time. For Augustine, time was the unexplainable, for the Greeks is was a god. Time is the motion of the hands of a clock, the sunrise in the mountains, the reality of train schedules, of hope, of waiting and forgetting. Time is "a tyrant that has his moods."[1] It rules over becoming and passing away, it connects earth with the heavens, today with tomorrow and yesterday. Clocks are its representatives in nature and human life. They are derived from a gestalt notion of time – astronomical models of the cosmos are also clocks – and lead to a continuum notion or the time's arrow notion of time. In Plato's astronomy, for instance, the "circle of the same" – the rotation of the heavens on its axis – represents pure periodicity (like the dial of an analogue clock), and the "circle of the diverse" – the motion of the planets on the ecliptic – represents a celestial calendar which makes it possible to count the days.[2]

Where time is ever present, it is subjected to analytic scrutiny in various ways. Different cognitive interests and different contexts determine different kinds of questions. Questions such us "Does time exist?," "Does time have properties such as a beginning and an end?" testify to an ontological interest and a metaphysical context. Questions such as "How is time felt?," "What role does time play for consciousness?" testify to a psychological interest and a context of theory of consciousness, of psychology or neurophysiology. Questions such us "What role is played by time in cultural and social connections?," "How are time and history related?" testify to the interest of social science or cultural studies and correspond to a social and cultural context. Questions such us "What does (natural) science know about time?," "How are the standards of time measurement formed?" testify to the interests of philosophy of science

1 J. W. von Goethe, *Gespraeche mit Eckermann, 25.2.1824*, in: E. Beutler (ed.), *Gedenkausgabe der Werke, Briefe und Gespraeche*, vol. XXIV, Zurich: Artemis 1948, p. 89.
2 See *Tim.* 37d-38e.

https://doi.org/10.1515/9783110596687-004

and to both an epistemological and a natural scientific context.[3] No wonder that Augustine despaired at the question "What is time?"

In the following, the discussion proceeds along a few natural scientific and anthropological pathways; the point of departure is an earlier, more detailed, analysis.[4] Philosophy has no exclusive access to the essence of time that could take the place of scientific or other approaches. However, in the form of epistemology and philosophy of science, it does possess critical instruments that may often still take hold where the scientific and the everyday understanding reach the limits set for them by theory and custom.

4.1 Arrows of time

The natural sciences, it seems, are bound to assign a prominent role to time. After all, lots of things change in the world; and change needs time to unfold. The reverse also holds true; as the saying goes, "the times they are a-changing." Thus, there appears to be ample use for time. This apparent evidence of the senses notwithstanding, the view has gained acceptance in some quarters that in reality there is no change. The true world is timeless. Becoming and passing away are mere illusions of the deceived mind and have no foundation in nature.

In order to address this issue in somewhat more detail, let us consider the physical basis for the "anisotropy" of time, one of time's most typical characteristics. This requires a careful investigation of the relevant theories of physics, a project prominently initiated by Hans Reichenbach[5] and Adolf Grünbaum[6], and later advanced by Huw Price[7], Lawrence S. Schulman[8], Heinz-Dieter Zeh[9] and others, to mention only a few.

---

3 See P. Janich, "Zeit," in: J. Mittelstrass (ed.), *Enzyklopaedie Philosophie und Wissenschaftstheorie*, vol. IV, Stuttgart and Weimar: J. B. Metzler 1996, pp. 827–831.

4 J. Mittelstrass, "From Time to Time: Remarks on the Difference between the Time of Nature and the Time of Man," in: J. Earman *et al.* (eds.), *Philosophical Problems of the Internal and External Worlds: Essays on the Philosophy of Adolf Grünbaum*, Pittsburgh Pa.: University of Pittsburgh Press 1993, pp. 83–101.

5 H. Reichenbach, *Philosophie der Raum-Zeit-Lehre*, Berlin and Leipzig: Walter de Gruyter 1928 (*The Philosophy of Space and Time*, New York: Dover Publications 1958).

6 A. Grünbaum, *Philosophical Problems of Space and Time*, 2nd edition, Dordrecht: Reidel 1973 (*Boston Studies in the Philosophy of Science*, vol. XII); "The Exclusion of Becoming from the Physical World," in: M. Čapek (ed.), *The Concepts of Space and Time: Their Structure and Their Development*, Dordrecht: Reidel 1976 (*Boston Studies in the Philosophy of Science*, vol. XXII).

7 H. Price, *Time's Arrow & Archimedes' Point: New Directions for the Physics of Time*, New York and Oxford: Oxford University Press 1996.

Clearly, the basis of physical time is to be sought in physical processes (as Aristotle already knew). The processes in question are *irreversible* in kind. Irreversible processes are temporally directed in that they go only one way; their temporal inverse does not occur. That is, no counter-processes exist that would be suited to restore the original type of state. In addition, we should distinguish between two kinds of irreversible processes. Processes are *nomologically* irreversible if the realization of the temporal inverse is excluded by a law of nature. They are *de facto* irreversible if their temporal inversion requires that particular initial or boundary conditions be realized which, as a matter of fact, do not occur.[10]

Grünbaum argues that the occurrence of *de facto* irreversible processes is sufficient to establish the anisotropy of physical time. *Anisotropy* means that there is a structural distinction between the two opposite time directions. If some states always follow one another in a fixed sequence, we are in a position to distinguish earlier and later states. Irreversible processes thus provide a basis for a physical implementation of the temporal relations "earlier than" or "later than," and this is precisely what the anisotropy of time comes down to. An isotropic time would require that no such physical implementation exists, which would in turn demand that all processes can actually be inverted. For that reason, the mere existence of irreversible processes, of whatever origin, confers anisotropy to time. And this is why the occurrence of *de facto* irreversibility is sufficient for anisotropy to emerge.[11]

What are the irreversible processes that might be suited to confer anisotropy to time? Consider processes such as the development of apparently homogeneous mixtures out of heterogeneous components, or the equalization of temperature differences. If cream is poured into a cup of coffee, both liquids mix as time passes; the reverse process, that is, the spontaneous separation of cream and coffee, has never been observed. Analogously, the differently heated ends of a metal rod will acquire the same temperature in the course of time; nobody has come across the spontaneous generation of a temperature gradient. Irreversible processes of this kind are described by the so-called *second law of thermodynamics*. According to that law, a quantity called *entropy* exists which, in every closed sys-

---

8 L. S. Schulman, *Time's Arrows and Quantum Measurement*, Cambridge: Cambridge University Press 1997.

9 H. D. Zeh, *The Physical Basis of the Direction of Time*, 3rd edition, Berlin: Springer 1999.

10 See A. Grünbaum, *Philosophical Problems of Space and Time*, pp. 209–210; "The Exclusion of Becoming from the Physical World," p. 474.

11 See A. Grünbaum, *Philosophical Problems of Space and Time*, pp. 211–212; "The Exclusion of Becoming from the Physical World," p. 475.

tem, either increases in time or, in the case of equilibrium systems, remains unaltered. The second law thus appears to express the existence of irreversible processes and to afford, consequently, the physical basis for the anisotropy of time. For every closed non-equilibrium system, higher entropy states are *later* than lower ones. In what follows, the focus will be on this thermodynamic basis of the anisotropy of time.

The seemingly smooth solution just outlined is, in fact, afflicted with a serious difficulty. Namely, the entropic behaviour of macroscopic thermodynamic systems should evidently be rooted in the behaviour of their microscopic constituents. It should be derivable from the motions of atoms or molecules. Ludwig Boltzmann attempted such a derivation and came up with his famous *H-theorem.* He applied the laws of mechanics to colliding gas molecules and assumed that the molecular motions proceed in a random fashion; that is, he supposed that on the mechanical level there are no preferred states of motion. On that basis, he arrived at a microscopically definable quantity that could be related to the macroscopic entropy, and which, correspondingly, exhibited the same temporally asymmetric behaviour. Since the randomness premise entering this deduction is, however, of an essentially statistical nature, the resulting version of the second law is likewise statistical in character. In contrast to its thermodynamic model, the mechanically derived second law does not rigorously preclude decreasing entropy values; it merely says that such cases are less probable and occur less frequently than the contrary cases of increasing entropy. Moreover, on the statistical variant, the entropy does not remain constant in equilibrium states but fluctuates irregularly around a fixed value. This implies that, in these states, the entropy drops and rises with equal frequency.

Grünbaum has shown, elaborating and substantially improving an earlier conception of Reichenbach's, how we can take advantage of this attenuated version of the second law so as to specify a physical basis for temporal asymmetry. It follows from the above considerations that we must resort to non-equilibrium systems for that purpose. Suitable systems of that kind are *branch systems.* Such systems branch off from their environment in a state of comparatively low entropy, remain closed for some time and finally merge again with the surrounding wider system. The low entropy state realized in branch systems is not the result of a statistical entropy fluctuation; rather, it is the product of an interaction with some outside agency of natural or human origin. Take an iced drink as an example. By immersing an ice cube into a lukewarm liquid we create such a low-entropy system. Subsequently, as the ice cube melts, the whole drink assumes a uniform temperature that finally agrees with the temperature of the environment. This process of continual equalization of temperature represents an increase in entropy with the result that the branch subsystem is eventually reintegrated into the larger system from which it was initially detached by the intervention. The temporal evolution of the entropy in branch systems is thus markedly asymmetric.

However, we have not yet taken into account the statistical nature of the second law. After all, due to that statistical nature it cannot be ruled out that some (untypical) branch system actually undergoes a decrease in entropy. This problem can be overcome by resorting to a large number of like branch systems. We employ a class of such systems characterized by the same initial macroscopic entropy states, and we *demand* that the microscopic realizations of these initial states be distributed at random. As a consequence of the second law, a typical representative of such a class will display a temporal increase in entropy. This feature constitutes the physical basis for the anisotropy of time.[12]

The next question to be addressed is: Why exactly does this temporally asymmetric behaviour of branch systems arise? The relevant point is the peculiar status of the irreversibility as expressed by the statistical variant of the second law. As already indicated, this law does not follow from the laws of mechanics alone. After all, the latter are perfectly reversible and do not exclude the time reversal of any process. This reversibility of the mechanical laws is reflected in the temporally symmetric entropic behaviour of equilibrium systems. What we need in addition, in order to arrive at irreversible processes, is reference to the prevailing *initial or boundary conditions*. The derivation of the statistical version of the second law requires premises that are not law-like in character but rather refer to actual circumstances as they *happen to occur*. Accordingly, the second law is not a fundamental law of nature; it arises from more basic, time-symmetric laws only by having recourse to particular states of affairs. This peculiarity is reflected in the branch-systems approach, for this method likewise relies essentially on the realization of particular non-equilibrium states that are supposed to exhibit microscopic randomness. Consequently, the anisotropy arising from this kind of irreversible processes is not nomological but merely *de facto*.

The conclusions drawn in philosophy of science often depend crucially on the particulars of the pertinent scientific theories. One exciting development in recent physics *might* induce a significant alteration as regards the philosophical interpretation of the directedness of time, which is the rise of irreversible thermodynamics. The relevant point here is the *status of the second law*. Underlying the more traditional picture outlined so far is the view that the laws of mechanics are more fundamental than the principles of thermodynamics. This is certainly not an implausible premise, since the thermal properties are supposed to arise

---

12 See A. Grünbaum, *Philosophical Problems of Space and Time*, pp. 254–260, for more detail.

from the statistical behaviour of microparticles whose motions are, in turn, governed by mechanical laws. It is precisely by way of this premise that the limited derivability of the second law from the laws of mechanics leads to the interpretation that the second law, along with the temporal anisotropy based on it, cannot be ascribed a fundamental status.

Since the development of the laser and the subsequent initiation of scientific research programmes such as synergetics, chaos theory, and irreversible thermodynamics, phenomena that occur far away from equilibrium have been of central interest. It turned out that order is often created far away from equilibrium. If we want to understand order, classical (equilibrium) thermodynamics is of not much help. This order consists in the fact that in appropriate systems, under conditions far from equilibrium, branchings of the system state – so-called bifurcations – occur. This means: If the relevant system parameters attain certain critical values, the system so to speak has the choice between two or more states; and under such circumstances it is minor fluctuations of the state of the system that determine the further development. Contrary to classical thermodynamics, minor fluctuations do not compensate for one another, rather they become macroscopically meaningful. This occurs by quasi-evolutionary mechanisms through which certain fluctuations are reinforced at the cost of others and thus determine macroscopic behaviour. The ordered structures that arise in this manner are based on small fluctuations; they are not determined by the initial conditions of the system. In this sense, *novel* structures arise in such processes. It seems that non-equilibrium thermodynamics allows "becoming" to enter nature. "The concept of order by fluctuations discards the static universe of dynamics for an open world, in which the new can arise through activity, in which development means innovation, creation and destruction, birth and death."[13] Becoming and passing away would now seem to acquire a central place in nature – within physical theory formation as well.

However, here too there is room for doubt. In spite of all the important and impressive theoretical and experimental progress of non-equilibrium thermodynamics during the past two decades, this field has changed nothing in the philosophical interpretation of time. On the contrary, even in the synergetic age the situation is precisely the same as, for instance, in Reichenbach's analysis of classical thermodynamics. In both, the analysis is based on irreversible processes under conditions of non-equilibrium. In addition for the philosophical interpretation of irreversibility, it is inconsequential whether these processes take place

---

13 I. Prigogine and I. Stengers, *Dialog mit der Natur: Neue Wege naturwissenschaftlichen Denkens*, 2nd edition, Munich and Zurich: Piper 1981, p. 204.

far from equilibrium (as in synergetics) or near to equilibrium (as in classical thermodynamics). The decisive point is that even in non-equilibrium thermodynamics, the anisotropy of time and the asymmetry of past and future are not anchored in the laws of nature, but are based on particular boundary conditions, on actually realized circumstances.

It turns out that the thermodynamic arrow of time is not time's only arrow. There are at least five other time-asymmetric processes, whose nomological relation to the thermodynamic arrow of time is still controversial.[14]

1. Radiation emitted by a source always expands in the future time direction. This is a well-known wave-phenomenon. Consider a stone that has been dropped in a lake; after a while one observes concentrically diverging waves on the surface of the lake. We never observe, however, waves that converge on a single point. From a theoretical point of view this is quite curious, since the basic equations of James Clerk Maxwell's theory permit such "advanced" solutions as well as the usual "retarded" solutions. Because of the lack of nomological constraints, one would expect, as in our discussion of the thermodynamic arrow of time, that the initial or boundary conditions of the universe are responsible for the radiating arrow of time.

2. According to our best cosmological models, based on Einstein's *general theory of relativity*, our universe is expanding. It expands in time, which in turn defines the cosmological arrow of time. Interesting questions occur when one tries to relate this arrow of time with the thermodynamic arrow of time. Thomas Gold[15] raised the question of whether the entropy rises or drops when the universe eventually recollapses in a "big crunch" as some models suggest. Given the second law, the entropy should of course increase, even if the universe returns to its initial state in the end. This initial state was, however, supposedly a state of very low entropy and can therefore only be reached again if the entropy decreases after a certain turning point. In order to resolve this riddle, ingenious arguments have been put forward, although a conclusive resolution has still not been reached.[16]

3. The *general theory of relativity* predicts that the gravitational collapse of a sufficiently massive star results in a *black hole*. After the collapse "the hole settles

---

14 See R. Penrose, "Singularities and Time-Asymmetry," in: St. W. Hawking and W. Israel (eds.), *General Relativity: An Einstein Centenary Survey*, Cambridge: Cambridge University Press 1979, pp. 581–638.

15 T. Gold, "The Arrow of Time," *American Journal of Physics* 30 (1962), pp. 403–410.

16 See C. Kiefer "Der Zeitbegriff in der Quantengravitation," *Philosophia Naturalis* 27 (1990), pp. 43–65; H. Price, *Time's Arrow & Archimedes' Point*; H. D. Zeh, *The Physical Basis of the Direction of Time*.

down and remains unchanging until the end of time."[17] The time-reverse of this process, which is permitted in principle, by the equations of the *general theory of relativity*, would be a singularity, known as a *white hole*, that sits for some indeterminate amount of time from the beginning of the universe and then erupts in a shower of ordinary matter. Given that we are now quite sure that black holes exist (presumably even at the center of our own galaxy, the Milky Way), without having any evidence whatsoever for white holes, there seems to be evidence for a *de facto* gravitational arrow of time which might, in turn, be connected with the cosmological arrow of time.

4. Microphysics also seems to have its arrow of time. In a quantum mechanical measurement process, a given state, expanded in a basis formed by eigenstates of the operator corresponding to the measured quantity, instantaneously collapses in one of these eigenstates with a certain probability, which can be calculated following the rules of quantum mechanics. This, at least, is the story told by the Copenhagen interpretation. Obviously, the measurement process is irreversible, because all the information about the other contributions to the state of the system prior to the measurement have been deleted. This information is lost – forever. That conclusion depends, however, on the interpretation of the quantum mechanical formalism one chooses. Take, for example, David Bohm's variant of quantum mechanics which turns out to be empirically indistinguishable from the standard formulation *à la* Niels Bohr and Werner Heisenberg. Given the close relation of Bohm's interpretation to classical statistical mechanics, it is clear that there is hope of eventually reducing the quantum mechanical arrow of time to the thermodynamic arrow – provided that the problem of the reduction of thermodynamics to statistical mechanics is solved.

5. Another arrow of time is quite curious. It only shows up once in a while in highenergy particle reactions that can be artificially created in the lab and which, perhaps, also showed up in the very early universe. Neutral K-mesons decay by violating parity and charge invariance. Due to the famous *CPT theorem*, decays of this kind therefore also violate time invariance. Unlike the other arrows of time presented so far, here we encounter, for the first time, a truly nomologically irreversible process. Unfortunately, this fundamental arrow of time resists any connection with, for example, the thermodynamic arrow of time.

Given this list of arrows of time, interesting foundational questions can be addressed. The most urgent one is perhaps how all these different arrows of time hang together. Is there a "master arrow" to which all other arrows of time reduce in some sense? Or do we have to accept a plurality of fundamental

---

17 R. Penrose, "Singularities and Time-Asymmetry," pp. 600–601.

and irreducible arrows of time in nature? A lot of work has been done on this issue by physicists and philosophers of science, and it is fair to say that so far no definite conclusions have been reached. There are several ingenious proposals to relate, say, all other arrows of time (with the possible exception of that involved in the decay of neutral K-mesons) to the cosmological arrow of time, but all have serious flaws, as for example Price[18] and others have shown in detail. We are therefore faced with a plurality of arrows of time in physics. Current physics does not provide us with a "master arrow", and it therefore also does not explain the direction of time.

However, maybe, this comes as no surprise since physics still lacks a final theory, which covers quantum mechanics (including quantum field theory) and general relativity. There are several proposals for such a *theory of everything* (such as supersymmetry), but due to the severe conceptual problems that these approaches face, a thoroughgoing discussion of the problem of the direction of time is quite difficult at the moment.[19] It turns out that the relevant fundamental equation, the socalled *Wheeler-de Witt equation*, does not exhibit a time variable at all. This suggests that we do indeed live in a kind of a Parmenidean block-universe, in which time is (at worst) an illusion of the human mind and (at best) an emergent phenomenon.[20] The hypothesis that time is not fundamental, but only shows up at a higher level of organization of the universe to order events, is currently favoured by most of the scientists working in this field. Many problems, however, remain. It has, for example, to be shown how this idea can be made more precise mathematically and conceptually. After all, time cannot emerge in time. So how and why does it emerge?

These are some of the problems discussed by physicists and philosophers of physics today. However, the concept of time has also facets which we experience in our everyday life.

---

18 *Time's Arrow & Archimedes' Point.*

19 See J. Butterfield and C. Isham for some recent work on time in the canonical approach to quantum gravity: "On the Emergence of Time in Quantum Gravity," in: J. Butterfield (ed.), *The Arguments of Time*, Oxford: Oxford University Press 1999, pp. 111–168. See also C. Kiefer, "Der Zeitbegriff in der Quantengravitation."

20 A moderate block universe view is defended by many philosophers of time, such as J. J. C. Smart, *Philosophy and Scientific Realism*, London: Routledge & Kegan Paul 1966, pp. 131–148, and H. Price, *Time's Arrow & Archimedes' Point.* According to this position, time is not something external to the world, but in it.

4.2 Time in the lifeworld

What is not in nature is *in us*. This is the philosophical alternative to a physical concept of time. Alongside physical time there enters, in the form of a "stream of consciousness," *experienced* time. This is the object not only of a theoretical interest, which also applies to the time concept of natural science, but also of an anthropological interest. The physics of time and the anthropology of time seem to be philosophical worlds apart. But this could be mistaken. If, physics too, as indicated, possesses no conclusive theory of time, then there might well not be so much distance between a natural scientific and an *anthropological* concept of time.

Augustine is taken to be the founder of the anthropological concept of time. For him, time was not a property of the world but a property of the soul. Time, understood as *lasting presence*, is not "measured" by motions in the world or in nature, but is "measured" by the *soul*.[21] Here a "cosmological" paradigm of time, such as Plato and Aristotle propounded, is replaced by a "psychological" or "mental" paradigm that is independent of clocks and replaces the concept of physical time by the concept of experienced time. No longer does the world, nature, (in the form, say, of periodic planetary motions) "have" time; only the soul "has" time. Thus, the soul is also the measure of time. That means that time has a *subjective* structure, and not just some time or other (i.e. the [physical] world has its time and man has his time or the soul has its time) but *the one and only time*. For Augustine, the memories of humans, their acts of bringing to mind (*memoria, contuitus, expectatio*), constitute time itself.[22]

The dilemma seems perfect: on the one hand, *physical* time, which as far as the essence of the laws of nature is concerned does not actually exist for modern physics or exists only in the boundary conditions of nature; on the other hand, *mental* time, which even the modern disciples of Augustine claim to be the only time. Between these two notions of time, epistemologically speaking: between a philosophical physicalism and a philosophical psychologism – time itself seems to lose its own proper reality. Where we expect time to be, somewhere between the time in nature and the time in us, there it is not, we are told by physics and philosophy.

To resolve this dilemma, the thesis is that temporal structures are taken from *action structures*. Neither nature nor the ego (the "soul") hold the key to time, but rather the way humans orient themselves in, and through, their actions. Actions

---

21 A. Augustinus, *Confessiones*, ed. J. Bernhart, Munich: Koesel 1955, pp. 632ff..
22 *Confessiones*, pp. 640ff..

occur not so much *in* time – this aspect is external to actions, for example, when we consider them relative to a time measured by a clock – rather they have their *own* time. That is, they have their own *duration* and their own *order*. Let us take chess as an example. Here a context of action has its duration and its order (in this case determined by rules). That the game of chess, as with other games, can also be played against the clock is rather an external arbitrary element. Games have their own time. Wherever they are adapted to the clock's time they lose a part of their essence.

Action contexts, such as chess, can be expressed in (measured) time. However, this only grasps a part of the actions that is *transferable* and thus is not a constitutive element of the action itself. In other words, every action – that is, every context of action – has so to speak two times, its *own* time and an *alien* time. Its own time is primary, the alien time secondary. If actions did not have their own times, we would not understand what time means. The "universal" or the *public*, i.e., social or cultural time has a derivative mode. It derives from the need – itself related to action – to compare and co-ordinate actions *and their times* (for instance in the form of train schedules or appointment calendars).

Nature too, that is, that which we perceive as "natural" time – for instance day and night, summer and winter – appears in this context primarily as the nature-like side of actions. Actions themselves do not distinguish between what in them is action and what in them is nature. However, actions can "in retrospect" be analyzed conceptually, for example, into a time of action and a time of nature, without identifying the former with the time of the "soul" and the latter with the time of physics. What is decisive is rather that we are dealing with a concept of time that does not suffer the previously mentioned dilemma between physicalism and psychologism and itself has a (temporarily) clock-free gestalt character.[23] Such gestalt-like forms of time are the time to go home, the time to say good bye, the time of love, and also the monsoon season, the age of Aquarius, and the strawberry season.

What was to be clarified about the concept of action can also be explained by the concept of *life*. Human time is reflected in the temporal gestalts of life. These are the "fateful forms" of life, because – like youth, maturity and old age – "they are essential to the process of development (...) being submits to them and endures them."[24] Youth and age, but also farewell and happiness,

---

23 The term *gestalt* is used here for the shape things take. In doing so, none of the holistic connotations often associated with the term in psychology are intended.

24 H. Plessner, *Die Stufen des Organischen und der Mensch: Einleitung in die philosophische Anthropologie. Gesammelte Schriften*, vol. IV, Frankfurt: Suhrkamp 1981, pp. 211–212.

are not properties of (individual) life but rather *forms*, *gestalts*, *times* that life submits to. In the "nature" of life lie the determinants of the time of life; life is not a temporal process. Or in another formulation, *the time of life is its times*.

Once again, what we have just considered has nothing to do with a physics of time that has lost sight of time, nor with an internalization of time that declares the soul to be the custodian of time. It is closer to Immanuel Kant's notion that time is a (pure) form or shape that underlies all our intuitions, experiences and actions.[25] We find time, which is our time, human time, neither in a four-dimensional manifold nor in the soul. This is how already the Greeks saw it.

When philosophy invented time as a philosophical problem, it was dealing with cosmological questions. Time, according to Plato, has a cosmic nature. It arose with the cosmos, whereby Plato distinguishes between the time forms of an ideal cosmos, called the "eternal animal" (ζῷον ἀίδιον[26]), a created cosmos, which also has an incorruptible structure, and the "moved" or measured time, which is the reflection of the astronomical nature of the cosmos, and like this has a periodic structure.[27] According to this notion, the cosmos is the totality of different forms of time, among them the forms of natural time, action time and life time. This is also clear in Aristotle: "[The cosmos] continues through its entire aeon unalterable and unmodified, living the best and most self-sufficient of lives. (...) for the fulfillment (τέλος) which includes the *time of life* of any creature outside of which no natural development can fall has been called its aeon."[28] That means, not only does the cosmos has its time, its aeon, but so does every living creature. But this can be understood along the lines of the assertion that time does not create life but life has its time.[29] This time is, at the same time, a cosmological "natural" time. Nature and time are still interwoven.

In Plato's conception this is expressed by saying that it is the "parts" of the moved time that "imitate" the structure of the whole of cosmic aeon. As the

---

25 In I. Kant's terminology (*Critique of Pure Reason* B 47) time is not a discursive concept but rather a "pure form of sensible intuition" (*Werke in sechs Baenden*, ed. W. Weischedel, Darmstadt: Wissenschaftliche Buchgesellschaft 1956–1964, vol. II, p. 79). That is, time is not an element of the physical world like heaven and earth; rather it has to do with our ways of experiencing and representing. Empirical time, that is, that which is the element of our experience, is formed in time which is not a part, an essential part, but the form of this experience. M. Heidegger later formulated this using the concept of the temporality of existence (*Sein und Zeit* [1927], Tuebingen: Max Niemeyer 1977, pp. 323 ff., 355 ff.).

26 *Tim.* 37d.

27 See *Tim.* 37d-38a; on the "astronomy" of time see *Tim.* 35a-36d.

28 *De cael.* A9.279a20–25.

29 Aristotle: its aeon; see G. Boehme, *Zeit und Zahl: Studien zur Zeittheorie bei Platon, Aristoteles, Leibniz und Kant*, Frankfurt: Klostermann 1974, pp. 80–81.

"time of the cosmos" has days and nights, months and years, so too human time has, for example, the periods of childhood, youth, adulthood, and old age. The life as well as the time of humans consists of structured wholes, that is, (temporal) gestalts. It has no temporal structure in the sense of dissolution in a temporal continuum or an arrow of time on which we are supposedly riding; it has purely temporal gestalts. Therefore, life consists not of time but of *times*, and these in turn "imitate" the temporal structure of the cosmos. Human time and human life represent cosmic time and cosmic life. It is senseless to mark boundaries for the temporal gestalts of life; the invention of the clock only leads us astray in this regard. An alternative would be a continuum model of time or the picture of time's arrow, but both of these belong to the world of physics. It is not *time* that flows but *things* that flow and change, albeit gestalt-like in time. Even Aristotle's theory of time does not allow such notions as are expressed in the phrase "time flies." For Aristotle, time is a form of motion.[30] However, in motion, time does not move; only *duration* grows analogously to the space traversed in a motion.[31]

Thus, the Platonic and the Aristotelian constructions of time correspond to the lived, and in mythical and everyday form represented time of their time. What might seem archaic here is rather the expression of lifeworld experience in dealing with human time, which has not yet made its way either into an abstract physical theory or into the soul. In other words, physics and anthropology have much to say to one another. That we seem today to have other experiences and that we look at things differently, lies in the need to co-ordinate concrete times. Modern constructions of time are doubtless theoretically superior to the notions of classical antiquity; but in lifeworld terms, from the point of view of human time, they are, in spite of their practical character, perhaps inferior. In this regard we are only misled by the fact that analogue watches that simulate the motion of the sun have become old-fashioned, and digital watches, which represent nothing but only "count," are on the march. At the same time, the insight of the Greeks, that life is not time, processual time, but rather has times, gestalt-like times, is fading. The difficulties that we today have with physical theories of time and with philosophical, psychological and social-scientific notions of the essence of time may lie in the fact that firstly (in a theoretical context), there is, obviously, more than one arrow of time, and that secondly (in a practi-

---

30 *Phys.* Δ11.219b23.

31 See P. Janich, *Die Protophysik der Zeit: Konstruktive Begruendung und Geschichte der Zeitmessung*, Frankfurt: Suhrkamp 1980, p. 255 (*Protophysics of Time: Constructive Foundation and History of Time Measurement*, Dordrecht: Reidel 1985, pp. 195–196).

cal context), we no longer live the gestalt-like character of time and no longer comprehend it.

5 Limits of Science?

The question about knowledge is, and ever was, also the question about what we do not know. Where there is knowledge, there is also ignorance. Ignorance thrives on knowledge, thrives on research. That is in a sense trivial. If there were nothing we did not know, knowledge, i.e. the search for knowledge, would lose its sense. Even the distinction between knowledge and ignorance could not be made. Likewise, the question whether there are limitations of knowledge and limits of science could not be asked. First of all, a few general remarks about knowing and not knowing, about limits of knowledge.

All of us have had the sobering experience of (really) knowing only a few things and not knowing a great many. All of us become painfully conscious of our intellectual limits both in private and professional affairs, that is, conscious of what we do not know and probably never will. Brains, it seems, have only a finite capacity to absorb, and life is short and for that reason alone unsuited to dreams of limitless knowledge – even ignoring the fact that one can imagine much more pleasant activities than stuffing oneself anew everyday with knowledge. Although Aristotle once said that all men, by their own nature, strive for knowledge,[1] he certainly did not mean the bookworm and the secluded scholar. Curiosity, which according to Aristotle is the form in which the human striving for knowledge is usually expressed, is more than scientific curiosity; it asserts itself in daily life, in travel, in experiencing the unusual and strange, in confronting closed doors and keyholes.

What is this peculiar feeling of confronting limits of knowledge, which unlike political and geographic boundaries are apparently not easy to cross? Is the occasional individual displeasure at knowing too little in the end the expression of a universal human incapacity, namely the incapacity to know everything, that is, to comprehend everything that exists in scientific form or even in non-scientific form?

Thinking this way already makes you half a philosopher. Indeed from the beginning it has been one of the favourite occupations of philosophy not only to ask about the conditions of origin of knowledge – how knowledge arises, what it presupposes – and about the essence of knowledge – what distinguishes knowledge from opinion, for instance –, but also about the limits of knowledge. Normally this meant limits of knowledge in the sense that the capacity and organization of the human understanding is simply not sufficient to answer all the questions that can be posed, nor to explore everything that can in any

1 *Met.* A1.980a21.

https://doi.org/10.1515/9783110596687-005

way be thought to be explorable. Human understanding is conceived as something like a lamp (or a search light) – knowledge reaches as far as its beam of light; beyond the reach of the beam lies darkness. Knowledge so to speak limits itself because the understanding is limited in its reach. And where it gropes about without light, it is blind, it fails to grasp its object and must abandon the field to others – mystics, esoterics, dreamers, all of them equipped with supposedly superhuman abilities. These businesses are booming again at the moment. An understanding that confronts its limits longs for the more simple, at least for the intellectual less taxing approaches to limitless knowledge – and the smaller the understanding the greater the longing.

But these are not the limits to be dealt with here or the limits of speculative or speculating knowledge, but rather the limits that really or perhaps only apparently arise for our *scientific* understanding. Indeed, the limits of knowledge are not only a theme for the everyday understanding or for philosophy but also for science itself, at least – now as reflection on the limits of science – at the point where science begins to become philosophical, that is, to view itself with philosophical eyes. This does not happen often, but it does occasionally happen. According to Paul Humphreys a limit is an "in-principle" epistemological constraint, whereas a limitation is an "in-practice" epistemological or pragmatic constraint.[2]

5.1 Completed science?

The second introductory remark – philosophy talking about limits of knowledge – may have sounded rather negative or at least regretful, as if we were obliged to submit to the inevitable. As a matter of fact, wherever we speak of the limits of knowledge, we become small as a rule, we experience our own limits, we draw anew the boundaries between the human situation and the divine. In the Middle Ages the limits of knowledge were always boundaries between the knowledge-seeking human and an omniscient God. Where the philosopher stopped, for instance, at the distinction between the natural and the supernatural, there the theologian began, and where he got stuck, for instance with the promise of beatific vision after death, there frolicked the mystics, magicians, ecstatics and

2 P. Humphreys, "Extending Ourselves," in: M. Carrier and G. J. Massey and L. Ruetsche (eds.), *Science at Century's End: Philosophical Questions on the Progress and Limits of Science*, Pittsburgh Pa.: University of Pittsburgh Press 2000, pp. 13–32.

elves. Even Faust, as we know, did not get much farther (and not just because of Gretchen).

Opposed to these notions of the limits of knowledge nourished by negative experiences or too-far-reaching reveries, there is a quite different notion that is even very positive, namely the notion of a *completion* or at least completability of knowledge. Limits of knowledge – that could also mean that knowledge comes to a standstill when *all is known*, when the world has given the human understanding knowledge of all that there is to know, when all corners have been searched, all riddles solved, all problems cracked, all Gordian knots cut and all things that seem impossible are provided with eggs of Columbus. Limits would lose their intimidation or their disappointment; they testify not to the failure of human understanding but rather to its triumph.

As a matter of fact, the history of science is full of expectations, prophecies and assurances in this direction. A few examples. In 1754 Denis Diderot, a brilliant writer, philosopher and (along with the mathematician d'Alembert) editor of the great French Encyclopedia (published from 1751 to 1780) wrote: I can almost assure you "that in Europe within the next century we shall not count three great mathematicians. This science will stand quite still where it was left by the Bernoullis, Euler, Maupertuis, Clairaut, Fontaine, and d'Alembert. They have erected the Pillars of Hercules. We shall not pass beyond them. Their works will continue through the centuries to come just like the Egyptian pyramids, whose stones, covered with hieroglyphics, invoke in us an awful idea of the power and resources of the men who erected them."[3] Mathematics and physics (the names listed stand for both) may come to be completed; science, like the pyramids, would become a grand exhibition piece visited by science tourists without a scientific future of their own.

American physicists in the 1970s expressed similar thoughts to the National Academy of Sciences: "It is possible to think of fundamental physics as eventually becoming complete. There is only one universe to investigate, and physics, unlike mathematics, cannot be indefinitely spun out purely by inventions of the mind. The logical relation of physics to chemistry and the other sciences it underlies is such that physics should be the first chapter to be completed. (...) Some unsolved problems might remain in the domain earlier characterized as organized complexity, but these would become the responsibility of the biophysicist or the astrophysicist. Basic physics would be complete; not only that, it would be manifestly complete, rather like the present state of Euclidean geom-

3 D. Diderot, *De l'interpretation de la nature* (1754), in: D. Diderot, *Œuvres complètes*, vols. I-XX, ed. J. Assézat and M. Tourneux, Paris: Garnier, 1875–1877, vol. II, p. 11.

etry."[4] In the same vein, David Lindley in 1994 spoke of the end of physics,[5] albeit with reference to the problem of a transformation of physics into a sort of l'art-pour-l'art mathematics. And Steven Weinberg says much the same: "My belief in a final theory rests (...) on the fact that our picture of nature has become ever more simple. (...) Of course, the mathematics has become more complicated, more difficult, more abstract. But the physical principles have become more elegant, more natural, and above all there are fewer of them. (...) And progress in the direction of simplicity must come to an end at some point."[6]

In this opinion, too, it is clear that the end of science is not the expression of incapacity in the sense of limits given to or forced upon human understanding; rather it can also be the expression of the achievement of the human understanding, indeed of its triumph over nature. Such limits of knowledge are filled-out boundaries; beyond these boundaries lies nothing that could be of scientific interest or which could present new tasks for science. America can only be discovered once (even though there might still be others who also consider themselves to be the discoverers), and the same holds, one would think, for nature. Once nature is discovered, once nature's laws are recognized and codified in textbooks, once everything has become simple in Weinberg's sense, there is nothing more to discover, and knowledge comes to a standstill. It can no longer be disappointed and replaced by better knowledge. It is no longer on the agenda of human understanding.

The extent to which such notions have gained currency was made clear by John Horgan, former editor of Scientific American, whose book "The End of Science" created a furore.[7] According to Horgan, the end of science is near in the sense that its success is its own end. According to him, all great discoveries have been made and all great theories written, all scientific questions answered; what remains is only scientific mopping-up, calculating decimal places, or didactic variations on a theme that can no longer be changed anymore. Science, in other words, is complete, and the human understanding has to look for new tasks and challenges beyond science. A peculiar notion.

---

4 A. Bromley *et al.*, *Physics in Perspective I*, Washington D.C.: National Academy of Sciences 1972, p. 80.

5 D. Lindley, *The End of Physics: The Myth of a Unified Theory*, New York: Basic Books 1993.

6 "Die Welt ist kalt und unpersoenlich: Physik-Nobelpreistraeger Steven Weinberg ueber den Traum von der Weltformel," *Der Spiegel*, No. 30, July 26, 1999, p. 192.

7 J. Horgan, *The End of Science: Facing the Limits of Knowledge in the Twilight of the Scientific Age*, Reading Mass.: Addison-Wesley 1996.

5.2 The sphere of knowledge

To the picture of a completed science, whose boundaries would be the boundaries of an impressive empire, we can oppose a different picture. It is the picture of a science that perhaps does not accord with the fantasies of the philosophers of science, who are infatuated sometimes with the infinite fallibility of science, sometimes with visions of its perfection, but it does accord rather well with the reality of science. This reality, known to every scientist, tells us that for every scientific problem that is solved new problems arise, that for every question answered new questions are posed, that for every insight gained a new ignorance is revealed. And that this ignorance does not merely refer to further decimal places is a very common experience of daily scientific work. We are dealing with questions like: How does a single cell develop into a complete organism? How did the galaxies arise, how did the whole universe arise? What goes on in black holes? What is (between philosophy and neuroscience) consciousness? One question succeeds the other and all of them are far from answered. We can visualize the problem in a picture that takes up a metaphor already used (in another context) by the philosopher and mathematician Blaise Pascal,[8] and employed again (in our context) by Herbert Spencer:[9] Scientific knowledge is a sphere floating in a space of ignorance and growing ever larger. As the sphere grows, its outer surface grows, too, and thus the surface of contact with ignorance grows.

This picture, which sees no limit to knowledge, can be interpreted in two different ways: a pessimistic version and an optimistic version. The *pessimistic* interpretation says (and here a small reminder from mathematics class may be useful but mercifully not absolutely necessary): If it is the *radius* of the sphere that represents knowledge, then, as the sphere increases in size, the area of the surface increases faster than the length of the radius, namely, as the second power. Thus ignorance grows faster than knowledge, or in other words: Scientific research produces a faster growth of ignorance than of knowledge. This is a star-

8 B. Pascal, *Œuvres complètes*, ed. L. Lafuma, Paris: Seuil 1963, pp. 525–528 (*Pensées* 199). Pascal refers to the cosmos here as an infinite sphere whose center is everywhere and whose periphery is nowhere. See also Herbert Spencer, who already uses the metaphor in the sense envisioned here (*First Principles* [1860], in: H. Spencer, *The Works*, vol. I, London: Macmillan 1904, p. 12). On the following, see J. Mittelstrass, "Die Wissenschaften und das Neue," in: J. Mittelstrass, *Leonardo-Welt: Ueber Wissenschaft, Forschung und Verantwortung*, Frankfurt: Suhrkamp 1992, pp. 74–95, here pp. 83–88 (I.4.3 Das Wissen und das Nicht-Wissen).

9 H. Spencer, *First Principles*, 5th edition, London and Edinburgh: Williams & Norgate 1890, pp. 16–17.

tling but – at least with regard to the elementary mathematics involved – correct result. In the *optimistic* version it is not the radius of the sphere that represents knowledge but its *volume*. As the sphere grows, its volume grows faster than its surface, namely as the third power of the radius. In this case (scientific) research still produces ever more ignorance, but knowledge is growing faster than ignorance.

Whichever interpretation of this sphere of knowledge one chooses, one thing is clear in this picture, and probably also in the experience of those scientists who do not deal immediately with the grand theories of their disciplines or with the grand design of all knowledge: The growth of knowledge does not make the world of the unknown, of the not-yet-explored any smaller, but rather larger. Research tasks grow with (growing) knowledge; there is no limit to the unknown. Rather it is the proximity, the constant contact to the unknown, sometimes – not only in the minds of philosophically minded scientists – the scent of the unthinkable or unimaginable (or what is taken to be so) that keeps science under its spell, that constitutes its stamina, even where what is already known seeks to present itself as knowledge on the brink of completion, or seems to be unreachable with the available means, or even seems to be inaccessible in principle for epistemological reasons. Here the limits are really dissolved, and (scientific) knowledge presents itself as essentially limitless. Neither are there boundaries with which the scientific understanding collides and gets a bloody nose, nor are there boundaries that form the contours of something completed. Is then the answer to the question "Are there limits to (scientific) knowledge?" a simple No?

5.3 Limits of science

Whoever gives a negative answer to the question as to the limits of (scientific) knowledge draws not only the vehement opposition of the proponents of the finitude thesis – the realm of the (scientifically) knowable is finite, therefore knowledge itself is finite –, he also opposes in a certain sense the everyday understanding and the philosophical understanding as well. For the everyday understanding, the existence of limits is something completely normal. Luck has its limits, skill in any area is limited by lack of skill at some point, alongside fulfilled wishes there are also unfulfilled ones. For the philosophical understanding, especially when it still sees itself as allied with the theological understanding, the existence of limits is something quite plausible and even alluring – where there are no limits, there is no reflection, no deep contemplation of limits;

and in such reflection and in such contemplation the philosophical understanding is particularly strong.

Historians of science see this too, in as much as they connect science with scientific progress. In a book on the idea of progress from 1932 we read: "Science has been advancing without interruption during the last three or four hundred years; every new discovery has led to new problems and new methods of solution, and opened up new fields for exploration. Hitherto men of science have not been compelled to halt, they have always found means to advance further. But what assurance have we that they will not one day come up against impassable barriers? The experience of four hundred years, in which the surface of nature has been successfully tapped, can hardly be said to warrant conclusions as to the prospect of operations extending over four hundred or four thousand centuries. Take biology or astronomy. How can we be sure that some day progress may not come to a dead pause, not because knowledge is exhausted, but because our resources for investigation are exhausted – because, for instance, scientific instruments have reached the limit of perfection beyond which it is demonstrably impossible to improve them, or because (in the case of astronomy) we come into the presence of forces of which, unlike gravitation, we have no terrestrial experience?"[10] Here limits in the negative sense appear, limits as unwanted limitations on (scientific) knowledge.

Philosophy of science discusses these questions, mainly in relation to the natural sciences, in the form of two theses:[11] (1) The thesis of the complete or asymptotic exhaustive survey of nature. This accords with the thesis of finitude, if we understand finitude as completion. According to this thesis, the history of scientific discoveries is either absolutely finite or at some point enters into an asymptotic approach to what can be known at all. The place of innovation would be then taken by mere elaboration and further precision. At some point science would have no future any more because, again, everything discoverable would have been discovered and everything would have been explained that was in need of a scientific explanation, and even the mopping-up operation, the calculation of further decimal places, the classification of additional cases that add nothing essentially new would gradually come to a close. (2) The thesis of the complete or asymptotic exhaustion of information capacities. This is the thesis that our historian of science propounded. According to this scenario, scientific information capacities are either absolutely finite or at some point enter into

---

10 J. Bury, *The Idea of Progress: An Inquiry into Its Origin and Growth*, New York: Dover Publications (1932) 1960, pp. 3–4.

11 See N. Rescher, *Scientific Progress: A Philosophical Essay on the Economics of Research in Natural Science*, Oxford: Blackwell 1978, pp. 6 ff..

an asymptotic approach to absolute information limits. Here, too, elaboration and further precision would replace innovation. Science would have exhausted its own possibilities for research and articulation and an "information barrier" would arise between science and nature, irrespective of whether investigation of the latter had reached a point of exhaustion. To overcome this barrier we must turn to science fiction – travelling with Star Trek and the space ship Enterprise off into scientific Wonderland.

The question, "Does (scientific) progress still have a future?" is thus only an apparent paradox. The question is, however, in fact unanswerable within the framework of the cited theses. This is evident, even without invoking once again the picture of the sphere of knowledge, from the connection between our research activity and our *goals*. If research is characterized not only by means of the states of research which have been achieved within individual disciplines (say, with regard to answering scientific questions) but also by means of the (internal and external) purposes associated with it (and this is doubtless the case, as a comparison of Newtonian physics with Aristotelian physics, each of which pursued completely different purposes, makes clear), then the notion of an end to scientific progress would include not only the assertion "We know everything (that we can know)" but also the assertion "We know all the purposes (that we can have)." The number of purposes, however, is unlimited, even if we take into account the limits to a scientific transformation of the world and of humankind. In other words: in order to answer the question, "Does (scientific) progress still have a future?" we would in some way already have to know what we do not know yet – what only progress or its failure to materialize could show us. In this sense then there are no limits to science. This applies also to the question of whether our sphere of knowledge rolls finitely or infinitely onward.

There is one thing that we can nonetheless know: our *resources* are finite. They are long since insufficient to provide science with what it needs in light of its growth and its wealth of ideas. There is also reason to believe that the higher the state of research reached, the more means must be applied to make equivalent advances (Nicholas Rescher calls this *Planck's principle of increasing expenditure*[12]). At one time a few silver coins were enough to move the world, even the scientific world; today it takes a large percentage of the gross national product of a country. Think of the construction and running of large particle accelerators in physics that devours huge sums in order to isolate one single further building block of the universe or to add further decimal places to measurements that

12 N. Rescher, *Scientific Progress*, pp. 79–94.

are considered significant or to confirm experimentally some constants of nature. Thus the discovery of the top quark, the last of the quarks according to the standard model of particle physics, demanded the employment of a group of 450 scientists from 35 institutes for about 20 years and devoured thousands of millions of Dollars. The question of marginal utility may be posed in science, too.[13] However, propositions and judgements of this kind depend not unproblematically on a *qualitative* and *quantitative* evaluation of (scientific) innovation. In the last analysis it is not the mass of scientific production, but the quality that counts. In any case it is true that progress limits itself if its costs grow faster than its results.

There are other limits that can be added, for instance, *ethical* limits. Such limits to scientific progress become evident today especially in the area of reproductive medicine and genetic engineering. The question is whether, driven forward by scientific progress, we are also *allowed* to do everything that we are *able* to do, for instance, intervening in the germ line of humans or cloning humans. Ethical limits to scientific (and technological) progress generally are drawn wherever these, instead of improving the living conditions of humans, turn against us, threatening and deforming us. A threat to or deformation of humans exists wherever humans are seen only as means not as persons, wherever, speaking with the German constitution, the inviolability of the dignity of man is no longer guaranteed. The physicist Max Born, who did foundational work in quantum theory, remembered: "In my youth it was still possible to be a scientist without paying much attention to the practical applications of science in technology. Today this is no longer possible; for natural science is inextricably entangled with social and political life. (...) Nowadays every scientist is a member of the technical and industrial system in which he lives. Therefore he must also carry part of the responsibility for a rational use of his results."[14] Albert Einstein was somewhat more sceptical on this score in 1948: "The tragedy of modern man lies in the fact that he has created conditions of existence for himself for which

13 On this and the distinction between positive cognitive limits (knowledge is complete) and negative cognitive limits (knowledge encounters insurmountable limits) see H. Tetens, "Kommt die Grundlagenforschung an ein Ende? Wissenschaftstheoretische Ueberlegungen zu den Grenzen der Wissenschaft," in: J. Mittelstrass (ed.), *Die Zukunft des Wissens. XVIII. Deutscher Kongress fuer Philosophie. Konstanz, 4.–8. Oktober 1999. Vortraege und Kolloquien*, Berlin: Akademie Verlag 2000, pp. 132–145.

14 M. Born, *Physics and Politics*, Edinburgh and London: Oliver and Boyd 1962, p. 63.

he is unprepared by his phylogenetic development."[15] What he means is that the impulses of the brain stem are stronger than the controls of the cerebrum.

In contrast to the (usually rather philosophical) question of the limits of knowledge, the question of the limits of scientific knowledge in particular cases is concrete, and thus, in the sense of real limits, it is also answerable. In the case of economic limits we are dealing with *factual* limits (there is not enough money); in the case of ethical limits we are dealing with *normative* limits (the obligatory and the permitted limit the possible). The fact that questions of economic and ethical limits are not always easy to answer depends on the fact that, in the case of economic limits, decisions as to priorities have to be made, and, in the case of ethical limits, we encounter a mixture of science and world view. Humans are not perfect beings, neither in regard to epistemology, nor with regard to economics and ethics.

An example of the question of ethical limits may help to clarify this, namely, the debate about cloning, in particular about the possible production of human clones (reproductive cloning).[16] This debate began with such a shock and has been pursued so hectically because the new potential of genetic technology has suddenly made possible something that once seemed to be forever beyond the horizons of human intervention, namely, the fabrication of a human being. It seems that boundaries drawn by nature itself are disappearing.

Producing clones means producing living creatures with the same genetic information, either by exchanging cell nuclei or by dividing embryos at very early stages of development. Cloning thus means that the genotype – that is, the primary hereditary material of two (or more) individuals – is the same, which does not mean that their phenotypes (the aggregate of external traits resulting from the genotype) are identical. Not all traits of an organism are wholly determined by the effects of the genes. The developmental conditions of an organism, including, in the case of humans, social and cultural conditions, also play an important role. In the case of identical (monozygous) twins, this has long been known. This makes it clear, by the way, that the production of clones is a thoroughly natural process; it is a replication mechanism that is quite common in nature, for instance, in bacteria and other microorganisms. What is new is only that this method can now be "artificially" applied to higher vertebrates. And what is of ethical significance is whether this kind of procedure may permissibly be applied to humans.

---

15 A. Einstein, *Ueber den Frieden: Weltordnung oder Weltuntergang?*, ed. O. Nathan and H. Norden, Bern: Herbert Lang 1975, p. 494 (my translation).

16 See J. Mittelstrass, "The Impact of the New Biology on Ethics," *European Review: Interdisciplinary Journal of the Academia Europaea* 7 (1999), pp. 277–283.

As a rule, especially in the arguments of theologians and philosophers, cloning humans is taken to be a severe infringement of human dignity, inasmuch as the *natural individuality* of humans is abrogated. On the one hand, an argument may be adduced appealing to the special relation in humans between two things, which in German are distinguished as *Leib* and *Koerper*,[17] that is, between the body as experienced phenomenologically and the body as a physical object. On the other hand, claims can be made about the character of humans as *ends in themselves*, who are to be protected from any kind of instrumentalization – such as the cloning of humans is taken to be. These are strong words that seem to be able to overturn any counterargument.

Nonetheless, it should first of all be recognized that there is an infringement of human dignity neither in the identity of the genomes of two people – identical twins are individual persons and bearers of human dignity – nor in the procedure of cloning itself. In this procedure, no person yet exists whose dignity could be attacked. On the contrary, as was made clear by a recent intervention, an infringement of human dignity occurs only through "the fact that a human being is produced as a means to an end that is not he himself, and that to this purpose, genetic identity with another human being is imposed on him."[18] This would be the case, for instance, in cloning for the purpose of producing donor organs or tissue – that is, establishing an individual organ bank. But this notion – the clone as a storehouse for spare parts – is absurd, since the clone, just as a natural twin, is of course an individual with all the rights that we associate with individuals. The fact that one (the clone) is just like the other (the cloned) is a circumstance that we have long been accustomed to in identical twins, whereby no one imagines that the one is (only) there for the other. Twins, too, are persons just like non-twins, and thus enjoy all the protections of the laws that enlightened societies afford to individualities.

There are, furthermore, a number of arguments that speak in favour of cloning or the affordance of such reproduction possibilities. What if the cloning pro-

---

17 On the basis for this distinction see H. Plessner, *Lachen und Weinen: Eine Untersuchung der Grenzen menschlichen Verhaltens* (1941, 3rd edition 1961), in: H. Plessner, *Philosophische Anthropologie. Lachen und Weinen. Das Laecheln. Anthropologie der Sinne*, ed. G. Dux, Frankfurt: S. Fischer 1970, pp. 11–171 (= *Gesammelte Schriften*, vol. VII [Ausdruck und menschliche Natur], Frankfurt: Suhrkamp 1982, pp. 201–398). According to Plessner, man *is* his phenomenological body (*Leib*) and *has* it as a physical body (*Koerper*); he is a "*Leib im Koerper*" as opposed to an animal that is its (physical) body and has it as its phenomenological body (*Gesammelte Schriften*, vol. VII, p. 238).

18 A. Eser *et al.*, "Klonierung beim Menschen: Biologische Grundlagen und ethisch-rechtliche Bewertung," *Jahrbuch fuer Wissenschaft und Ethik*, vol. II, Berlin and New York: Walter de Gruyter 1997, pp. 357–373.

cedure is used as a method of treating infertility or is applied to avoid serious hereditary illnesses? Even a widow's wish for a child very much like the one she has just lost[19] might be a permissible reason for applying the cloning procedure. In such cases, there is an infringement neither of the principle of the inviolability of human dignity nor of the closely connected determination of man as an end in himself, as formulated for instance in Immanuel Kant's second form of the *categorical imperative*. Remember that even this formulation of ends in themselves is both realistic and humane: "Act so that you use humanity, whether in your own person or in the person of any other, always at the same time as an end, never merely as a means."[20] If we use Kant in our arguments we should read Kant closely. He argues here that we must not treat a person *merely* as a means but always *also* as an end. Kant intends no complete exclusion of the means perspective here. Had he said: never under any circumstances as a means, then every instance of human reproduction would be morally reprehensible, because it is always, as is the act that leads to it, not only determined by the person as a purpose. The progenitors of a child think not only of the happiness of the child but also of their own. In another formulation: It would be completely unrealistic to assert that up to now the only thought at the conception of children has been the happiness of the future child, and not for instance the happiness of the parents or compensation for the loss of an earlier child.[21]

Thus it is clear that – whatever apparently powerful arguments have been pulled out of the arsenal of philosophers and theologians[22] – cloning itself is not in any sense "in itself" reprehensible, but only in connection to particular human intentions. However, the question of principle remains. How much technology do we want to place in the stead of traditional modes of behaviour that are considered natural? After all, with the technology of cloning we change not only future generations, but we also change ourselves, at least in our self-understandings. In other words: Wherever boundaries are crossed, which, as in the case of human reproduction, seem to be set by nature, we must analyse very pre-

---

19 See Ph. Kitcher, *The Lives to Come: The Genetic Revolution and Human Possiblities*, New York: Simon & Schuster 1997, p. 336.

20 I. Kant, *Grundlegung zur Metaphysik der Sitten* (1785), in: I. Kant, *Gesammelte Schriften*, ed. Koeniglich-Preussische Akademie der Wissenschaften (zu Berlin), Berlin: Georg Reimer 1902ff., vol. IV, p. 429.

21 See C. F. Gethmann, "Ethische Argumente gegen das Klonieren von Menschen," Europaeische Akademie zur Erforschung von Folgen wissenschaftlich-technischer Entwicklungen Bad Neuenahr-Ahrweiler GmbH, *Academy Letter*, No. 9 (4/1998), p. 2.

22 See D. Birnbacher, "Die Fortpflanzung hat ihre Unschuld verloren: Ein Gespraech," *Information Philosophie*, No. 3 (August 1998), p. 114.

cisely and without recourse to individual intuitions or to ideological prejudices just where such boundaries ought to lie in the future. For man is a being without measure that can only live by means of measures. To find these measures and thus to set the right limits is difficult even when, as in the case of our example, they seem clear to many people, for whatever reasons.

5.4 Productive incompleteness

If the reflections presented here are correct, or expressed somewhat more modestly, if they do not seem implausible, then the question, "Are there limits to knowledge?" formulated here as the question, "Are there limits to science?" can actually be answered, in the ethical and economic sense, with a clear Yes (though in the ethical case this is not always easy), and in the epistemological case with just as clear a No – possibly limited by the consideration that only the progress of science itself or its failure to progress can really show whether the scientific understanding in fact has (internal or external) limits.

This answer is confirmed by epistemological efforts concerned with demonstrating epistemic finitude. For example, in discussions concerning the boundaries of science and the limits of our knowledge, reference is often made to Kurt Goedel's incompleteness proof, to undecidable theorems and to Werner Heisenberg's so-called *uncertainty principle.* Heisenberg's principle, in opposition to classical physics, states that there are quantum theoretical laws according to which arbitrarily precise and simultaneous measurements of certain states of a system are impossible (more precisely, of canonically conjugated magnitudes, for instance the position and momentum of a particle). Thus we have here a lawlike limit on the boundaries of the precision measurement of certain physical magnitudes. On the other hand, according to Goedel's theorem[23] there is in each consistent logical system strong enough to permit the construction of arithmetic at least one sentence which is true, but which cannot be derived within the system. Furthermore, the assumed consistency of the system cannot be demonstrated by the means provided by the system itself (taken some natural constraints as granted). Goedel's theorem thus points not only to boundaries on that which can effectively be achieved, but also to an unexpected distinction between truth and derivability. The same holds in the case of the related undecidability theorems, according to which a sentence *A* is undecidable if there is no

23 See B. Buldt, "Unvollstaendigkeitssatz," in: J. Mittelstrass (ed.), *Enzyklopaedie Philosophie und Wissenschaftstheorie*, vol. IV, Stuttgart and Weimar: J. B. Metzler 1996, pp. 432–436.

general finite algorithm, i. e. neither a semantic nor a constructive algorithm, according to which it can be decided whether *A* is true or not[24] (which, by the way, holds also for the weaker notion of derivability, for derivability is undecidable for most systems beyond propositional logic). These proofs thereby set bounds on effective procedures.

Now in all of these cases, that is in the incompleteness and undecidability theorems of logic and mathematics as well as in the so-called uncertainty principle of physics, we have affair less with limits to science or with limitations of our knowledge, than with results of the scientific construction of knowledge. More precisely: we are dealing with scientific results concerning the limitation of the knowable, in so far as the latter is based on effective procedures. That is to say, theorems such as those of incompleteness or of undecidability do not set limits on knowledge in the conventional sense. They are instead part of logical and metamathematical knowledge. And thus they cannot be used against this knowledge, i. e., they cannot be employed as arguments against the power or capacity of this knowledge.

The same holds with regard to their application in other contexts, for instance in the area of the philosophy of mind or the mind-body problem. Thus Alfred Gierer, for example, attempts to found his thesis concerning theoretical limits on our understanding of the relation between mind and body with recourse to Goedel's theorem and a principle of finitism.[25] Gierer holds the monistic position that brain- and consciousness-states are identical, but that the exact relationship between the two kinds of states cannot be explained. Goedel's theorem is supposed to support the thesis, that the brain cannot develop a complete representation of itself. Thus, according to Gierer, one is to expect that undecidability problems will arise in the context of abstract statements concerning brain-processes, or in the context of reflexive statements concerning such processes (for instance concerning self-consciousness).[26]

What appears in Gierer's development (by means of an appeal to a computer analogy, which is to reveal the limits of a decoding of the mind-body problem) as a limit to knowledge, is revealed on closer examination to be merely a limitation dictated by the choice of method, in this case the use of the computer analogy and its connection to Goedelian incompleteness.[27] The argument is furthermore

---

24 See K. Lorenz, "Wahrheit," in: J. Mittelstrass (ed.), *Enzyklopaedie Philosophie und Wissenschaftstheorie*, vol. IV, pp. 582–587.

25 A. Gierer, *Die Physik, das Leben und die Seele*, 3rd edition, Munich and Zurich: Piper 1986.

26 A. Gierer, *Die Physik, das Leben und die Seele*, pp. 46, 237, 247.

27 See M. Carrier and J. Mittelstrass, *Mind, Brain, Behavior: The Mind-Body Problem and the Philosophy of Psychology*, Berlin and New York: Walter de Gruyter 1991, pp. 251–257.

questionable even on these methodological assumptions. For Goedel's theorems have as an essential presupposition that one is dealing with a consistent formal system. In consequence one would have to show, first that the brain-consciousness system can be represented as a formal deductive system, thus is nothing more than a Turing-machine, second, that this system is consistent, and third, that the resulting formally undecidable sentences have a relevant status or meaning. But this cannot be done. Furthermore, we are dealing here with a repetition of an old question of Kant's, a question which must be answered, as it was by Kant himself, in the negative, namely the question of whether the subject can cognize itself. This question can also be formulated as: Whether or not consciousness, or any cognitive system, can give a complete description of itself. For Kant, this question has two sides. On the one hand, the unity of consciousness, the *transcendental ego*, is not knowledge of itself in a consciousness-theoretical sense. On the other hand, the subject cognizes itself in empirical self-cognition only to the extent that it is available to itself in intuition.[28] The question of whether or not the subject, whether or not consciousness or some other cognitive system can describe itself completely is thus directed less toward some suspected limitation on knowledge than at its very openness.

Besides, scientific thought is so to speak constantly re-inventing itself, realising itself in its constructions and destroying itself with its constructions. The *phoenix* is the symbol of science just as the *owl* is the symbol of philosophy. Science creates itself, just as philosophy considers itself and what it has seen. Science thrives on the mortality of knowledge, philosophy on the immortality (or better limitlessness) of reflection, which therefore constantly encounters itself, while science forgets and discovers. Only the concept of construction holds the two, philosophy and science together. For philosophical reflection, too – so long as it does not just reproduce itself hermeneutically – constructs, devises new worlds, only to fill them again with its age-old experiences.

Those who find these statements too speculative, too philosophical, may prefer the more modest formulation, which is however equivalent in content, that scientific knowledge must as a rule be taken to be imperfect or incomplete and incorrect, though not in the sense of a defect – such a notion would in fact presuppose an attainable perfection or completeness – but in the sense of an openness of scientific knowledge in principle. Furthermore, paradoxically formulated, the boundlessness of science, in the sense of an interminable progress of knowledge, lies precisely in the limited character of knowledge, in its finitude and corrigibility.

28 See I. Kant, *Critique of Pure Reason* B 157, B 408.

Here I would like once more to invoke the sphere of knowledge. It floats in a space of ignorance and grows ever larger, thus increasing its surface of contact with ignorance. Or in another formulation: The limits of science are either *error limits* – the scientific understanding gets stuck in its own insufficiency – or *economic limits* – scientific progress becomes unaffordable – or *moral limits*, which are ever more often given in cases where scientific progress is directed against humankind itself. In any case, any standard or measure of science, also that which sets limits to its progress (in the realm of economics or ethics) is a *practical* one, not a *theoretical* one. Errors turn out to be normal, resources turn out to be finite, (ethical) norms turn out to be compelling, although this is not due to any insurmountable properties of the knowable or of the knowing subject's understanding. This means that science does indeed have practical limits, but no theoretical limits. And this equates the human and the divine in science.

6 Transdisciplinarity

The concept of *transdisciplinarity* – which in the context of the philosophy of science was introduced 30 years ago, as a further development of the concept of *interdisciplinarity*[1] – has found a foothold in science and is even becoming fashionable. It is used not just by science, when thinking about its own research practice, but also by science policy, when trying to give the impression of being knowledgeable in the philosophy of science. More and more often it seems as if transdisciplinarity were self-explanatory, as if its meaning were evident. But this is not at all the case. Though there are attempts to define transdisciplinarity as a method in order to present an elaborated methodology to the sciences, this is rather due to a misunderstanding than to an insight, namely the misunderstanding that transdisciplinarity is something amenable to a formulation in a theoretical form. More on this later. The first question concerns the relation of what we label transdisciplinarity to the disciplinary structure of science. To put it differently: Does disciplinarity, which has accompanied us on our scientific roads, have a future? And is interdisciplinarity, often evoked when addressing the good relations between the disciplines among themselves, no longer enough? What, in any case, is transdisciplinarity, and how could its institutionalization look like?[2]

6.1 Disciplinarity, interdisciplinarity, and the new complexity of science

The scientific system has become complex in a worrying manner. This is not just valid for the ever-increasing acceleration of the growth of knowledge, but also in organizational and institutional respects. A particularization of subject matters and disciplines is increasing; the capacity to think in disciplinarities, that is, in larger units of science, is decreasing. The borders of subjects and disciplines, if they are still perceived as such at all, threaten to turn into limits not just of institutions, but also of discovery. Accordingly, the concept of interdisciplinarity,

---

1 J. Mittelstrass, "Die Stunde der Interdisziplinaritaet?," in: J. Kocka (ed.), *Interdisziplinaritaet: Praxis – Herausforderung – Ideologie*, Frankfurt: Suhrkamp 1987, pp. 152–158.

2 See also J. Mittelstrass, *Transdisziplinaritaet – wissenschaftliche Zukunft und institutionelle Wirklichkeit*, Konstanz: Universitaetsverlag Konstanz 2003 (*Konstanzer Universitätsreden*, vol. 214).

https://doi.org/10.1515/9783110596687-006

often used to oppose this development, is being viewed as a repair tool, which, as time goes by, is supposed to lead to a new scientific order.

But interdisciplinarity is neither something normal, nor something really new, nor the true essence of the scientific order. Where it works, it rectifies misguided developments of science, but also renders apparent that (scientific) thinking in larger disciplinary units has manifestly declined. A whole should again arise out of particularities, both in a systematic as well as in an institutional sense. In what follows, the institutional aspect, aiming at the rebuilding of genuine disciplinarities, will not be addressed primarily, but the role of structures and strategies that span subjects and disciplines in research (and, mediately, also in teaching).

In the first instance it is advisable to remind oneself that subjects and disciplines have grown through the history of science, and that their boundaries are thus determined neither by their objects themselves, nor by theory, but by historical growth. Furthermore, their identity is determined by certain objects of research, theories, methods, aims of research, which often do not correspond univocally to the definitions of subjects or disciplines, but which instead overlap these disciplines. This does not just become apparent in the fact that disciplines are being guided by methodical and theoretic ideas which, as with the concepts of a law of nature, of causality, and of explanation, are not determined to belong to any one discipline, but also in the fact that those problems to find solutions for science serves, often do not fit straightforwardly into a disciplinary framework. For instance, the disciplines dealing with the theoretical description of heat were by no means the same in the history of this problem. Initially, heat was conceived of as the inner movement of matter, and thus as an object of physics. In the theory of caloric matter, formulated by Hermann Boerhaave at the beginning of the 18th century, and later developed by Antoine Laurent de Lavoisier, heat, conceived as matter, becomes an object of chemistry. Finally, with the kinetic theory of heat, heat changes disciplines anew and becomes an object of physics again. So not (just) the objects define the discipline, but our manner of dealing with them in theory.

The example may also be generalized so as to say that certain problems cannot be captured by a single discipline. This is true, in particular, of those problems, as for instance rendered clear in the fields of environment, energy and health, which arise from issues not exclusively scientific. There is, and this not just in these fields, an asymmetry in the developments of problems and scientific disciplines, and this is aggravated as the developments of disciplines and science in general are characterized by an increasing specialization. But this means that the interdisciplinarity appealed to in this situation is not a ritual of fashion, but arises from constraints deriving from the problems themselves.

If the problems, whether scientific or not, do not do us the favour of defining themselves in the terms of a particular discipline or subfield, then special efforts will have to be undertaken, which will normally take us outside our normal subjects or disciplines. In other words, irrespective of the sense in which interdisciplinarity is being understood here, as interdisciplinarity reconstructing larger disciplinary perspectives, or as a real enlargement of the domain of interest of the scientific fields and disciplines, or going beyond scientific fields and disciplines, one thing should be clear: interdisciplinarity, understood rightly, is not merely an alternation between the disciplines, nor is it hovering over them, like Hegel's absolute spirit. Rather, it undoes disciplinary rigidities whenever these obstruct the formation of problems and corresponding research-based actions; in reality, then, it is *transdisciplinarity.*

6.2 Transdisciplinarity

Whereas scientific cooperation in general means the readiness to engage in cooperation in science, and interdisciplinarity normally means concrete cooperation with a finite duration, transdisciplinarity is intended to imply that cooperation will lead to an enduring and systematic scientific order that will change the outlook of subject matters and disciplines. Transdisciplinarity is a form of scientific work which arises in cases concerning the solution of non-scientific problems, for instance the above mentioned environmental, energy and health care policy problems, as well as an intrascientific principle concerning the order of scientific knowledge and scientific research itself. In both cases, transdisciplinarity is a *principle of research and science*, one which becomes operative wherever it is impossible to define or attempt to solve problems within the boundaries of subjects or disciplines, or where one goes beyond such definitions.

Besides, pure forms of transdisciplinarity occur equally rarely as do pure forms of disciplinarity. These, too, mostly conceive and realize themselves in the context of neighbouring scientific forms, for instance with sociological elements in the work of the historian, chemical elements in the work of the biologist or biological elements in the work of the medical researcher. In this respect, disciplinarity and transdisciplinarity are research-guiding principles or ideal types of scientific work, but mixed forms are the rule. What is important is that science and research be aware of this, and that productive research not be restricted by concerns that are obsolete (and mostly simply due to habit), and thereby focused on narrow areas. Such restrictions neither serve scientific progress, nor a world which, in light of its own problems, wants to use rather than admire science.

In other words, transdisciplinarity overcomes the narrow areas of subjects and disciplines which have been constituted historically, but which have lost their historical memory and their problem-solving capacities due to an excessive specialization. But it does not lead to new disciplines. That is why it cannot replace fields and disciplines. Transdisciplinarity, secondly, is a scientific principle of work and organization which spans subjects and disciplines, driven by specific problems, but it is not transscientific. The optics of transdisciplinarity is scientific, and it is directed at a world that is more than ever a work of the scientific and technical mind, and which has a scientific and technical nature. Thirdly, transdisciplinarity is a principle of research, and not, or at most mediately, namely when the theories themselves follow transdisciplinary research programmes, a theoretical principle. It guides the perception of problems, and their solution, but it does not solidify in theoretical forms. That is why transdisciplinarity is not a method, or even elaborated in the form of a methodology.

What might still appear very abstract has already found its concrete form in scientific practice, and it is increasingly being fostered institutionally. This applies, for instance, to new scientific centres which have been formed in the USA, in Berkeley, Chicago, Harvard, Princeton and Stanford[3], for instance in Harvard the *Center for Imaging and Mesoscale Structures*. It addresses a range of issues which could not sensibly be attributed to any particular discipline. Their object of research are structures of a certain dimension in general, not any particular objects. Other institutional forms are conceivable, even without gathering them in one building, such as, for instance, in the case of the *Center for Nanoscience (CeNS)* at the University of Munich.

Such Centres are also no longer organized according to the traditional pattern of faculties or schools of physics, chemistry, or biology, but rather according to a transdisciplinary perspective, which, in this case, follows the actual developments of science. That is also true in cases where single problems are being addressed, as for instance in the new *Bio-X* Center in Stanford[4], or the *Center for Genomics and Proteomics* in Harvard.[5] Biologists here use sophisticated methods from physics and chemistry to find out about the structure of biologically relevant macromolecules, and physicists like the Nobel Prize winner Steven Chu, one of the initiators of the *Bio-X*-programme, investigate biological objects

3 See L. Garwin, "US Universities Create Bridges between Physics and Biology," *Nature* 397 (January 7, 1999), p. 3.
4 See L. Garwin, *ibid.*
5 See D. Malakoff, "Genomic, Nanotech Centers Open: $ 200 Million Push by Harvard," *Science* 283 (January 29, 1999), pp. 610–611.

which may be manipulated with the most advanced methods from physics.[6] Competencies acquired in individual disciplines remain a fundamental precondition for tasks defined transdisciplinarily, but they no longer suffice to successfully tackle research projects which extend beyond the established fields. This will, in the future, lead to new organizational forms, also besides the establishment of centres such as those mentioned, in which the boundaries between the individual fields and disciplines will fade away.

And this is true of all institutional forms of research and science, not just of research undertaken in universities. In Germany, these constitute a highly differentiated system, which ranges from university research, defined by the unity of research and teaching, and Max-Planck-research, defined by ground-breaking projects in the newest areas of science, to large-scale research, defined by big machinery and fixed-term projects of research and developments (once upon a time openly declared as being of national interest), and Fraunhofer-research, defined by applied research and closeness to industry, to research done in industry, defined by a close connection between research and development.

But the logic of this system, which other countries envy not just because it demonstrates scientific rationality but also exceptional efficiency, is starting to become problematic. That is because it leads to the evolution of independent subsystems whereas really – in the spirit of the above mentioned development of centres – the formation of connections at a low level should be the name of the game, not the expansion of independent systems at a high institutional level. For Germany, but certainly also for other countries, this means that institutionalized research networks of limited duration should take the place of subsystems of science which are isolating themselves more and more from each other. The justification for this is simple, especially from the perspective of science: *The system of science has to move when research is moving.* At the moment, things are rather exactly the opposite. It is not the research that finds its order, but an order which is given in its subsystems and getting increasingly solidified is looking for suitable research. This order of science is becoming contraproductive. And this should not be the future of research and of a system of science such as the German. As may be seen, the increasing transdisciplinarity of scientific research will, or should, have far-reaching institutional consequences.

6 See L. Garwin, *ibid.*

6.3 The Unity of Nature

In the course of the development of modern science, ideas of a *unity of nature* are again gaining philosophical and scientific importance, as a view of a unitary physical theory – if there is only one nature, then all natural laws must be part of a unitary theory of nature[7] –, but also as research is increasingly taking a transdisciplinary perspective. If nature does not distinguish between physics, chemistry and biology, then why should the sciences that research it do so, let alone by means of a rigid disciplinary framework? Indeed, the original idea of a unity of nature is shining through the transdisciplinary orientation of modern research programmes. But this idea shall not be the theme here. Instead, the point to be considered is how transdisciplinarity is, as a matter of fact, not just a philosophical dream, but instead a part, even an essential part, of the latest scientific research. Two examples.

6.3.1 Nanotechnology

The idea to do research on and reconstruct functional structures of the dimension of 10^{-9} and 10^{-6} metres, that is, individual atoms, molecules and small collections of atoms, originates with a visionary talk given in 1959 by the physicist and later Nobel Prize winner Richard P. Feynman at a conference of the American Physical Society (APS) at the California Institute of Technology in Pasadena.[8] In this talk, Feynman addressed, among other things, the storing and reading of information on very small spaces – and so anticipated a number of currently used lithographic methods –, miniature computers and small "artificial surgeons" which would move through blood vessels to do their job there, or in the heart. Feynman had been inspired, as he says himself, by biology, in which such small and highly functional structures may already be found. Why should it not be possible to create them artificially?

Nanotechnologists examine extremely small functional structures (normally biological), for instance membranes, enzymes and other cellular components (*wet nanotechnology*) and try, furthermore, to experimentally create these structures, using, for instance, semi-conductors (*dry nanotechnology*) or to simulate their properties on computers (*computational nanotechnology*). In the creation

7 See C. F. von Weizsaecker, *Die Einheit der Natur: Studien*, Munich: Hanser 1971.

8 R. Feynman, "There's Plenty of Room at the Bottom," *Engineering and Science* 23 (1960), pp. 22–36.

of nanostructures, scientists from physics as well as chemistry work together closely. Whereas physicists usually begin with a given structure, for instance a surface, which they then process with methods from physics (top-down approach), scientists from chemistry start at the level of atoms and molecules to systematically assemble them (bottom-up approach). All areas of nanotech research are closely interrelated; advances in one area normally entail advances in other areas.

Among the most significant advances in nanotechnology are the synthesis of carbon rings (*fullerenes*), the creation of microscopic tubes made out of carbon atoms[9] and the man-made concatenation of a very small number of carbon atoms.[10] It is remarkable that it is the biologically important carbon atom that gets used as "raw material."

6.3.2 The quantum-mechanic measurement process and the concept of information

There are some questions and areas of research whose results are not clearly attributable to physics or philosophy. The quantum-mechanical measurement process is one of these. How is it possible that the measurement on a quantum-mechanical system leads to a definite and unambiguous result even if the state measured has been prepared as a superposition of eigenstates of the measured observables? Does the wave function collapse instantaneously, at the moment of measurement, into one of the eigenstates contained in the superposition (as the adherents of the Copenhagen interpretation maintain)? Or do we perceive only a part of the true wave function after the measurement (as for instance the many-worlds and the many-minds interpretations suggest)? Or is the measurement process a "real" process, occurring on an extremely small timescale, whose genuinely non-linear stochastic dynamics goes far beyond the basic assumptions of quantum mechanics and, strictly speaking, even contradicts them?[11] Other questions pertain to the unifiability of quantum mechanics with the *theory of special relativity*[12] and

---

9 See P. M. Ajayan and T. W. Ebbesen, "Nanometre-Size Tubes of Carbon," *Reports on Progress in Physics* 60 (1997), pp. 1025–1062.
10 See R. A. Broglia, "Wires of Seven Atoms – Feynman's Very, Very Small World," *Contemporary Physics* 39 (1998), pp. 371–376.
11 See G. C. Ghirardi and A. Rimini and T. Weber, "Unified Dynamics for Microscopic and Macroscopic Systems," *Physical Review* D 34 (1986), pp. 470–491.
12 See T. Maudlin, *Quantum Non-Locality and Relativity: Metaphysical Intimations of Modern Physics*, Oxford and Cambridge Mass.: Blackwell 1994 (*Aristotelian Society Series*, vol. 13).

the role of non-local causality in physics. Philosophers, wellacquainted with subtle differences and the handling of concepts in need of explanation (in this context, for instance, non-locality), prove to be useful partners of physics, not always – the will to conceptual clarity often shows itself to be a weak will –, but occasionally. (Of course, it is true that philosophy should not want to solve problems which science, as here physics, is in a better position to solve.)

Information technology too proves to be useful here. According to the Copenhagen interpretation, a quantum system loses information when a measurement is being executed on it. The reason is that the system is in a state of superposition just before the measurement is performed, whereas after the measurement it takes, qua projection, the eigenstate of that operator that has been assigned to the observable to be measured. Any further information of the original state has been lost. After introducing the concept of information into quantum mechanics, the theory of information may be used to further analyse the measurement process, so that a bridge to further applications in technology has been built (for instance, "quantum cryptography" and "quantum computing"[13]). The research principle of transdisciplinarity does not just concern the collaboration of diverse scientific skills, it also extends to technology.

Do these examples together with what was said about transdisciplinarity before, imply that we are facing a fundamental paradigm change, in which it is not the theoretical concepts that change – as in the transfer from Aristotelian to Newtonian physics – but in which the order of our scientific knowledge, and thus that of our scientific research and education, is changing fundamentally? It will not get that far, for reasons already mentioned while explaining the concept of transdisciplinarity. The standards of rationality, and with them the methods and forms of theory construction, are not changing. It is the forms of organization of science and research which are doing so. Once again, transdisciplinarity is a *principle of research and science*, which applies wherever a definition of problems or solutions just through individual fields or disciplines is not possible, or goes beyond them. It is not a *theoretical principle* that changes our textbooks. Just like competence in particular fields or disciplines, transdisciplinarity is a research-guiding principle and a form of scientific organization, but with the peculiarity that transdisciplinarity removes narrowness due to specialization which is due not to scientific necessities but to institutional habits.

13 See H. Weinfurter and A. Zeilinger, "Informationsuebertragung und Informationsverarbeitung in der Quantenwelt," *Physikalische Blaetter* 52 (1996), No. 3, pp. 219–224.

ciplines, formation of interdisciplinary competence, de-disciplinarization in the argumentative, transdisciplinarity as argumentative unity. What is crucial is the argumentative perspective, the condition that the entire process took place, in a non-trivial sense, in argumentative space. In this example: the unity looked for, the determination of health care standards and the determination of measures for a good life, had been created going beyond, as well through different disciplines.

In other words, the methodical in this practical transdisciplinarity consists in its argumentative creation and the steps which may be distinguished in the process of creation. This again may also apply to the transdisciplinarity previously described as theoretical, or inner-scientific. That too bases itself on disciplinary competencies, but does not relate them to objects of the disciplines, and thus constitutes a new "disciplinarity," which, with respect to the original disciplines, becomes transdisciplinarity. A research programme, for instance the structural research mentioned before, goes beyond the common disciplinary determinations, and develops its own forms of work and thus also changes the disciplines involved (also due to the constitution of the problem to be solved). That means: Within the boundaries of transdisciplinary developments, the individual disciplines do not remain what they were, at least, they change their methodical and theoretical perspectives. Not just the theories in the narrow sense, also the disciplines themselves get pulled into the process of research and science – in a systematic manner. Precisely this is what is meant with methodical transdisciplinarity.

7 Pragmatic Dualism in the Philosophy of Mind

Consciousness and self-consciousness, or self-understanding, are among the central concepts of philosophy in its European tradition – like nature and reason. Man is the animal which is conscious of its doings, its cognition and its situation in the world and which is able to relate, at the same time, to this consciousness cognitively and reflectively. Philosophy addresses these relationships in the domain of epistemology, but increasingly so, too, does natural science in the form of (cognitive) neuroscience and, in particular, brain research. The natural sciences are getting involved with philosophical conceptions, but philosophy is getting equally involved with scientific procedures and results. This latter proceeds by way of the philosophy of science (of the neurosciences), as well as by way of more anthropological approaches. Knowledge about man is scientific and philosophical (epistemological and anthropological) at the same time. This sometimes gives rise to conflicts, especially when scientific knowledge claims to encompass all knowledge of man. Everything that is the case is amenable to scientific explanation – thus the fundamental conviction of the natural sciences. Is this also the case with consciousness and self-consciousness?

As far as the natural sciences are concerned, the objective is to explain how consciousness works from the physiological point of view and what capacities it has – in the words of the brain researcher: "to attribute a large part of our cognitive and motoric capacities to the brain and to conceive of deficiencies of these functions as entirely standard organic diseases."[1] As far as philosophy is concerned, the objective is to explain from the epistemological point of view how consciousness is mirrored in its cognition and its other subjective performances. The cognition and the reflection of this mirroring, in turn, is self-consciousness. The natural sciences and philosophy are also at loggerheads about this topic, self-consciousness. Is it possible to "explain" self-consciousness just like it is possible to explain consciousness, or is it of a different kind? The natural sciences say No, philosophy says Yes, and tries to express this in the conceptions used – for instance the concept of the *self* or the *ego*. Let us start with a short reminder of the career of the concepts of consciousness and self-consciousness.

---

1 W. Singer, "Einfuehrung," in: *Gehirn und Bewusstsein*, Heidelberg and Berlin and Oxford: Spektrum Akademischer Verlag 1994, p. VII (my translation).

https://doi.org/10.1515/9783110596687-007

7.1 The *ego*

Consciousness has always been understood as a cognition that is not just displayed in mere behaviour but which articulates itself as a "consciousness of something." To articulate here means: to differentiate, to conceptualize, to assess, to connect the perceived (what is given in perception) with the constructed (what is determined in thought). In the process consciousness becomes, in modern terminology, a property of the mental or of mental states and conditions. This finds its epistemological expression in the Aristotelian concept of thinking, *noesis*, complementing a mere perception with an aspect of intentionality: consciousness as an action directed towards a certain matter of fact or as object-related knowing, which articulates itself linguistically (conceptually).

In contrast to the concept of consciousness as object-related perception, the concept of self-consciousness means the perception of an object-related perception (and other subjective performances), a consciousness, thus, which becomes self-reflective and to this extent also may be understood as condition of all cognition in its philosophic and scientific forms. Also this aspect may already be found in Aristotle, namely in the phrase "thinking of thinking" (νόησις νοήσεως)[2], where Aristotle assigns the concept of a pure self-consciousness to the concept of a pure reason, which turns out to be a condition of philosophy and science.[3] In René Descartes, this issue becomes the fundamental principle of his philosophy of science and metaphysics; it also does this in the further development of both rationalist and empiricist epistemological perspectives. In Gottfried Wilhelm Leibniz, for instance, the perceptions of the (rational) monads (souls) are apperceptions, defined as reflective consciousness, in John Locke, the "ideas of reflection" are the result of perceiving one's own cognitions. At the same time, the concept of self-consciousness is related to the concept of an *I-substance*, the *ego*, which in Kant – where all currents of philosophical tradition meet and are put on new, critical foundations – in turn finds its transcendental reformulation.[4]

---

2 *Met.* Λ9.1074b34.

3 On this and the further history of the concept of self-consciousness, see C. F. Gethmann, "Selbstbewusstsein," in: J. Mittelstrass (ed.), *Enzyklopaedie Philosophie und Wissenschaftstheorie*, vol. III, Stuttgart and Weimar: J. B. Metzler 1995, pp. 755–759.

4 The following is closely based on an earlier account: J. Mittelstrass, "Le soi philosophique et l'identité de la rationalité philosophique," in: E. D. Carosella *et al.* (eds.), *L'identité changeante de l'individu: La constante construction du Soi*, Paris: L'Harmattan 2008, pp. 203–212 (especially pp. 207–210). English version: "The Philosophical Self and the Identity of Philosophical Ration-

In Kant's terminology, the *ego*, in reference to itself, perceives itself as *appearance*, not as substance, and it is empirically given only in this sense. With the concept of the *transcendental subject*, this idea gains epistemological significance as the principle of unity of knowledge and things: "For inner experience in general and its possibility, or perception in general and its relation to another perception, without any particular distinction or empirical determination being given in it, cannot be regarded as empirical cognition, but must be regarded as cognition of the empirical in general, and belongs to the investigation of the possibility of every experience, which is of course transcendental."[5] Self-consciousness, in this sense, again means the ability of the subject to refer, with the intention of knowing, to its own object-related knowing. Kant uses here the well-known formula "The *I think* must *be able* to accompany all my representations", followed by the explanation: "for otherwise something would be represented in me that could not be thought at all, which is as much as to say that the representation would either be impossible or else at least would be nothing for me."[6] From here the further development leads on the one hand, against the idealistic theory of the *ego* and self-consciousness, to a philosophy of concrete subjectivity, to a phenomenology of *ego*-perceptions and, on the other hand, to psychological theories as well as analytical approaches.

In this development, the concept of the *self* is either identical with the concept of the *ego* or, in contrast to this concept, emphasizes the more phenomenological aspects of individual forms of existence and self-understanding. For Leibniz, it was self-reflection that makes it possible to say "*I*," Kant distinguishes between a *determining self* (thought) and a *determinable self* (the thinking subject) without associating this distinction with any distinctions between *ego* and *self*. In contrast to the identity of the *ego* by which more abstract aspects are emphasized – these aspects still influenced Edmund Husserl's concept of the transcendental *ego* –, the concept of the *self*, for instance in Martin Heidegger, aims at the phenomenal variety of personal identity (the "authentically existing self,"[7] "*Dasein*" [existence] as "being-within-the-world").

As already pointed out, the concept of reflection is closely related to the concepts of the *ego* and the *self* or rather the concept of self-consciousness. This con-

---

ality," in: J. Chr. Heilinger *et al.* (eds.), *Individualitaet und Selbstbestimmung*, Berlin: Akademie Verlag 2009, pp. 55–61 (especially pp. 58–59).

5 *Critique of Pure Reason* B 401 (translation from *Critique of Pure Reason* [transl. and ed. by P. Guyer and A. W. Wood], Cambridge: Cambridge University Press 1998, p. 412).

6 *Critique of Pure Reason* B 132–133 (translation from *Critique of Pure Reason* [see footnote 5], p. 246).

7 *Sein und Zeit*, 8th edition, Tuebingen: Niemeyer 1957, p. 130.

cept stands for the self-ascertainment of the *ego* or the *self*, in epistemological terms for the "*I think*" which, according to Kant, accompanies all judgements and all activities of the understanding. Thought, in this respect, is self-reflexive by nature and, correspondingly, so is the cogitating *ego* and the cogitating *self*.

A further step is made by Johann Gottlieb Fichte when he says that in thinking the *ego* creates itself. Here, the *ego* is perceived as absolute *ego*, as pure self-performance that even constitutes in itself the difference between *ego* and non-*ego*, i.e. nature. With this, a logical level is reached where it is no longer the constitution of the individual that is at stake (according to the related concepts of *ego*-identity and *self*-identity) but, as in Kant in an epistemological framework, the constitution of a philosophical *ego* or philosophical *self* – in Kant's terminology: the constitution of a transcendental subject. The identity of this subject consists in the fact that it is neither the particular (empirical) subject nor the universal (theoretical) subject, but the condition of both. In this sense, Ludwig Wittgenstein writes: "The subject does not belong to the world, but it is a limit of the world."[8] Wittgenstein here refers to the individual subject, but his statement is also precisely true in view of the fact that the acting *ego* (Kant: the determining *ego*), in its performances, cannot be grasped theoretically.[9] Just this is expressed in the concepts of reflectivity and the transcendental.

What is expressed by the terminology of *ego* or *self*, as well as in the expression that the acting or determining *ego* cannot be grasped theoretically, marks the frontier at which the natural sciences in the figure of brain research and philosophy in the form of epistemology and ethics may stand – and sometimes do stand – opposed to one another in critical conflict.

7.2 Science and the philosophy of mind

Where science claims to explain everything or at least a great variety of many different things with the same method, it either becomes dogmatic or lets itself be guided by a methodological and theoretical paradigm that makes a claim to universality. The kind of physicalism propounded in the context of Logical Empiricism is an example. It says that all knowing may be expressed using the language of physics and, what is more, that all scientific theories are ultimately reducible to theories of physics. This is an expression of the covert or open

---

8 *Tractatus logico-philosophicus* 5.632 (translation from: *Tractatus logico-philosophicus. With an Introduction by Bertrand Russell*, London: Kegan Paul, Trench, Trubner & Co. 1922, 1947, p. 151).
9 See K. Lorenz, "Identitaet," in: J. Mittelstrass (ed.), *Enzyklopaedie Philosophie und Wissenschaftstheorie*, 2nd edition, vol. III, Stuttgart and Weimar: J. B. Metzler 2008, pp. 530–534.

reductionism of the natural sciences, that is, the programmatic idea of tracing back scientific explanations to uniform notions of a conceptual, methodological and theoretical kind, aiming at a universal explanatory competence. Philosophically, in the traditional sense, this is a variety of *monism*, which in this case leads to a *naturalism*. Naturalism claims that scientific claims of validity are to be traced back to natural facts (that is, facts captured by science), which implies a naturalizing of cognition itself. In the neurosciences, especially in brain research, such a claim is based on the thesis, that also characterizes physicalism, of the causal closure of the physical world. That would include, accordingly, also the spheres of consciousness and self-consciousness.

In modern philosophy, more specifically in the philosophy of mind, this corresponds to the so-called *eliminative materialism* and the so-called *identity-theory*, especially in the theoretical variety of type-identity. According to *eliminative materialism*, cognitive psychology and folk psychology will be replaced, materially and conceptually, by progress in neurophysiology[10]; according to the *identity theory*, mental states and processes are identical to states and processes of the human brain.[11] Following the theory of type-identity, this includes the claim of a (future) reducibility of psychological statements to neurophysiological laws.[12] Thus *eliminative materialism* and the *identity theory* represent the philosophical foundations of the reductionist claims of (parts of) brain research, to be (or to become) the "whole" explanation in matters of consciousness and self-consciousness.

By no means, however, have all issues been resolved, as far as science or philosophy is concerned. In fact, philosophy of mind leads to a trilemma, which has been represented by the following three theses: "*Radical diversity:* mental phenomena, that is, the mental states, processes or events which we experience, are not physical. In other words, they are *strictly different* from all physical phenomena. *Mental causation:* Mental phenomena may *cause* physical phe-

---

10 See P. M. Churchland, "Eliminative Materialism and the Propositional Attitudes," *The Journal of Philosophy* 78 (1981), pp. 67–90; P. S. Churchland, *Neurophilosophy: Toward a Unified Science of the Mind/Brain*, Cambridge Mass. and London: MIT Press 1986, 1988.

11 See H. Feigl, *The "Mental" and the "Physical". The Essay [1958] and a Postscript*, Minneapolis Minn.: University of Minnesota Press 1967; J. J. C. Smart, "The Mind/Brain Identity Theory," in: *The Stanford Encyclopedia of Philosophy (Fall 2011 Edition)*, E. N. Zalta (ed.), URL = <http://plato.stanford.edu/archives/fall2011/entries/mind-identity/>.

12 For an account and discussion of various theories in the domain of the philosophy of mind see M. Carrier and J. Mittelstrass, *Mind, Brain, Behavior: The Mind-Body Problem and the Philosophy of Psychology*, Berlin and New York: Walter de Gruyter 1991; also M. Carrier, "philosophy of mind," in: J. Mittelstrass (ed.), *Enzyklopaedie Philosophie und Wissenschaftstheorie*, 2nd edition, vol. III, pp. 220–226.

nomena, that is, our conscious intentions may cause bodily movements in the external world. *Causal closure:* The domain of physical phenomena is causally closed, that is, physical states, processes and events have only *physical* but no non-physical causes."[13]

The third thesis is the thesis adopted by large parts of the cognitive neurosciences. It corresponds, in the philosophy of mind, to the *identity theory* understood as a theory of type-identity and thus philosophically represents a reductionist and naturalist worldview. But it is built on sand. In fact, the cognitive neurosciences, and in particular the brain sciences, have not yet managed to demonstrate causal (neuronal) mechanisms which could explain consciousness, let alone self-consciousness – and thus the interactions of brain and mind. "Consciousness is and remains mysterious."[14] By making reference to the thesis of causal closure of nature, this position proves to be unfounded from the scientific point of view, and so does the thesis of determinism it endorses – there is no uniform principle of causation in modern physics which could serve as a foundation for a strict determinism – and from the philosophical point of view it proves to be metaphysical.[15] For instance in quantum theory, probabilistic state descriptions of micro-objects lead to indeterminist theories; in philosophy, metaphysical points of view give way to conceptions from the philosophy of science and language. But thus the trilemma mentioned above is losing its philosophical significance: the first and the second thesis remain philosophically viable, the third does not. It is the language of the cognitive neurosciences in particular that give the false impression of neuronal determinism and thus a worldview which seemingly does not leave space for the distinction between physical and non-physical phenomena anymore.[16]

---

13 B. Falkenburg, *Mythos Determinismus: Wieviel erklaert uns die Hirnforschung?*, Berlin and Heidelberg: Springer 2012, p. 28–29 (my translation).

14 B. Falkenburg, *Mythos Determinismus*, p. 379.

15 See B. Falkenburg, *Mythos Determinismus*, pp. 370–378.

16 On this and for a critique of this worldview, see M. R. Bennett and P. M. St. Hacker, *Philosophical Foundations of Neuroscience*, Malden Mass. and Oxford and Carlton: Blackwell 2003; also P. Janich, *Kein neues Menschenbild: Zur Sprache der Hirnforschung*, Frankfurt: Suhrkamp 2009. Janich draws attention to the consequences of an alleged neuronal determinism, namely the obligation to attribute sense and reference to neuronal states and processes themselves: "When a brain researcher makes the claim that 'ultimately' the meaning and validity of linguistic communication should be explicable via neuronal functions, which already have meaning and validity, he is fudging. He is cheating his way to acceptance of his claim by already attributing the properties of meaningful speech to the material building blocks of his models of the brain" (*Kein neues Menschenbild*, p. 73 [my translation]).

Mediating between the positions of strict neuronal determinism and that of metaphysical dualism as supported by the rationalist tradition, but also, for instance, by Karl R. Popper and John C. Eccles[17] (the independent existence of mental and physical states), today there also is the conception of *pragmatic dualism.*[18] This view leaves the theoretical possibility of body-mind identity open, but indicates that psychophysical interactionism might be the most convincing option, given the current state of science, also avoiding the above-mentioned trilemma. "Consciousness," "self-consciousness," and "*ego*" are dualistic terms; they cannot be formed in a monistic conception. They are titles of a specifically philosophical way of orientating oneself in thought and through thought without blocking the way to science. That means: the conception of pragmatic dualism neither anticipates future scientific developments nor does it exclude any particular scientific developments, nor does it simply adopt uncritically earlier metaphysical positions developed in the context of the so-called mind-body problem.

7.3 Free will

Dualist and monist views of consciousness and self-consciousness clash nowhere as vehemently as they do on the question of *free will.* Not just for philosophical reason, but also for common sense, conscious decisions are the causes of actions. First we decide, then we act; first there is consciousness, then there is the action. In opposition to that stands the thesis of the neurophysiologist that consciousness is a merely *interpreting* and not an acting authority: it is not consciousness that moves, other, physical and mental circumstances move. On that view the actual causes of actions are connected to physical and psychological mechanisms, which are not amenable to introspection, in which consciousness is looking at itself, as it were, and thus not amenable to conscious experience. But this would mean that consciousness would not have a privileged access to the originating conditions of an action; instead it would find itself in the position of a third-person spectator, as it were. It is not influencing decisions, but only registering them and dressing them in a meaning that appears plausible. It invents good reasons, which however do not have anything to do with the actual causes. It is quite clear that this conception appears rather perplexing, considered against the background of our self-experience and our self-understanding.

17 K. R. Popper and J. C. Eccles, *The Self and Its Brain*, Berlin and London and New York: Springer 1977, 1985.

18 M. Carrier and J. Mittelstrass, *Mind, Brain, Behavior* (footnote 12).

How could we be able to imagine that consciousness, experienced subjectively as the source of our activities should be causally ineffective.

Already for Greek thought, the concept of the will marks the passage from prudence or deliberation to action; the freedom of the will accordingly meant the space of action between doing and not doing something. The issue was the concept of well-founded willing, not the search after some mysterious substance in body, soul or reason. A person whose actions are guided by rational considerations is free, or rightly called the possessor of a free will. The further philosophical development went a different way. The will is now considered a separate source of action, next to reason. By leaving behind the prudence model of the will problems of determinism arise for the first time. They do not concern the idea of free action but rather the idea of free willing, that is, the idea of the free will as an uncaused willing. The thesis is: We may do or not do what we will; but we cannot will or not will whatever we will. This is how Arthur Schopenhauer's writings and his thesis of the world as will and representation is to be read.

The right keyword gets voiced (as so often is the case) with Kant. Next to a "causality according to the laws of nature" there is a "causality of freedom." The issue is again (just like in Greek thought) *freedom of action*, not some sort of substance, called freedom or free will, and the problem of well-founded (rational) action, in traditional terminology: the problem of a rational (or good) will. "Causality of freedom" – this is, in other words, the capacity to act according to *principles.* The point is demands (in the sense of principles) addressed to ourselves, and the realization of these demands. Everybody, including the natural scientist, understands what is meant by this, even if it is a "causality according to the laws of nature" and not a "causality of freedom" that he is looking for as scientist. Using the terminology of freedom and the will: We are free in formulating the principles and in (willingly) following them. It is not the free will that is the problem, but the rational will (and thus the determination of the will as self-determination), articulated in the demand to act according to rational reasons.[19]

When people see this differently and take the causal closure of nature for granted, as natural scientists do, it is primarily *semantic* problems that cloud the view on the differences. In this particular case, the scientific side is unable to imagine anything else than that the non-scientific positions supported by others will eventually join them in believing that the decision between "(free will)

19 See J. Mittelstrass, "Der arme Wille: Zur Leidensgeschichte des Willens in der Philosophie," in: H. Heckhausen *et al.* (eds.), *Jenseits des Rubikon: Der Wille in den Humanwissenschaften*, Berlin etc.: Springer 1987, pp. 33–48, also in: J. Mittelstrass, *Der Flug der Eule: Von der Vernunft der Wissenschaft und der Aufgabe der Philosophie*, Frankfurt: Suhrkamp 1989, pp. 142–163.

exists" and "(free will) does not exist" will be made on experimental grounds, so to speak, that semantic problems are mere pseudo-problems. But the untenable position of causal closure and the arguments brought forward for pragmatic dualism render it obvious that they are not pseudo-problems. It is, in any case, also true that if the claim that the will is causally determined throughout were true, then that claim itself and its claim to validity would be determined causally, that is, via natural causalities. Or to put it differently: A world without freedom would be a world without *reasons*, and for this reason – this is often overlooked by the reductionist and naturalist approaches – a world without science. Hence: science itself is the most beautiful refutation of the scientific negation of a free will.[20]

So, too, in the question of the freedom of the will it is important not to overshoot the scientific target, namely the explanation of physical and mental phenomena, in the direction of the unity of the physical world, and to take semantic matters seriously. Philosophy should take account of scientific developments, but science should also acknowledge philosophical distinctions. It appears that the conception of pragmatic dualism provides the best basis for this.

---

20 Also the Libet experiments, according to which 300 milliseconds before a conscious "act of the will" takes place, the corresponding readiness potential may already be measured, only yield the desired conclusion, that we are not free in our decisions but determined by natural causalities if the muscle contraction taking place after the readiness potential has built up may be interpreted as an act of the will or expression of such an act. But precisely this needs to be also justified.

II Nature and History of Science

8 From Plato's World to Einstein's World

Nature is not a simple concept – for at least three reasons. First of all, nature, meaning the physical universe, is in itself an evolving nature. It is neither fixed nor always the same. There is biological evolution (genetic diversity and variation) and – derivatively (though the concept is problematical) – cosmic evolution, e. g. stellar evolution. Therefore, concepts of nature follow nature in its evolution or ought to follow it. Second, nature has, at least in a historical perspective, different meanings in different cultures. For example, the Greek tradition distinguishes between creative nature (*natura naturans*) and created nature (*natura naturata*), the Indian tradition identifies nature and earth and speaks of the Goddess Earth. Today, under the influence of the modern sciences, historically divergent concepts of nature have lost their scientific significance. Nature is now what is governed by universal laws, although a universal determinism is limited in many ways by the occurrence of probabilities, i.e. by probabilistic laws. Third, though initially nature was just that part of the world that man had not made, it has now, to a great extent, become part of an artificial world built by science and technology.

This makes it difficult to distinguish clearly between what is natural and what is not. For example, what is the exact meaning of the concept of nature for the particle physicist, who, so to speak, creates his objects in big machines, or for the molecular biologist, who rearranges genomes? Is it still nature that scientists investigate and humanists reflect on when they speak about nature and culture and the cultural impact of science and technology on nature? It is not only that different cultures have generated different concepts of nature, but also that science interferes with nature in a way that makes it often difficult, even with respect to the concept of universal laws of nature or probabilistic laws, to give a determinate definition of what nature is. The old duality between nature and culture as well as the duality between universalism or determinism and probabilism has been superseded by a plurality of new dualities, among which is again also the duality between a creative or evolving nature and a created nature (now in a scientific framework).

Closely connected with changing concepts of nature are *world views* or *world pictures*. World pictures are models of the reality, depending on particular *concepts*, like the concept of nature, *beliefs* about how the reality functions, and *programmes*, like scientific programmes. It is not only mythical cultures which create world pictures, science, too, generates them. It provides the world with a picture in which it appears as what it ostensibly is "in itself," as "nature," as "evolution," as "creation," regardless of how we transfer the conception of a

https://doi.org/10.1515/9783110596687-008

"world outside us" to "our" world, the world in which we live. To an understanding of science also belongs an understanding of its power to constitute the world and to generate world pictures, particularly with respect to nature. In the following, some examples are given of how science pictures nature and the world – from Plato's and Aristotle's world to Einstein's and Heisenberg's world.[1]

With Plato's world, i.e. with Plato's cosmological concept, the idea of a philosophical as well as a scientific cosmology is born. Here, in Plato's dialogue *Timaios*, a powerful craftsman creates the world according to a perfect model, namely the "cosmos" of the Platonic ideas. Like a perfect living being, the cosmos turns out to be an animated rational being, as a visible god in the form of a perfect sphere. Its soul, the "world soul," has an astronomical nature: it is formed by the mathematical order of the trajectories of the planets. At the same time the planets function as "tools of time;" time (χρόνος), arising with the heavens, is an image of eternity (αἰών). The planets are visible and created gods, the earth the "most venerable goddess in the heavens." Man in the cosmos, which consists of purely godlike entities and is itself a living god, is compared with a plant, which roots "not in the earth but in the heavens," he connects the earth with the heavens related to him. Later on, in Christian thought, i.e. in Christian platonism, the world of Platonic ideas to which the craftsman refers as a perfect model, becomes the realm of thoughts of God creating the world.

Unlike a *Plato world*, which, apart from the mythical language in which it is presented, is governed by mathematical (geometrical) and astronomical laws, Aristotle's world is a world of natural things that consist of matter and form and have within themselves a source of motion. Motions caused by such a "natural" source are "teleological" motions, i.e. they make a thing into what, according to its own nature, it really is, or they lead it, in the form of a "natural" local motion, to its "natural" place. A theory of natural positions, incorporated in a theory of elements, corresponds in this sense to a theory of simple (natural) bodies (bodies that have a source of motion in themselves) and simple motion (the motion of simple bodies). In the cosmological dimension, an *Aristotle world* consists of eleven spheres grouped around the central body, earth. Each such sphere is constituted by two concentric spherical surfaces: the three inner spheres housing the elements and the eight outer spheres housing the then known planets

1 This is a short and revised version of an earlier contribution which also dealt with questions of the history and philosophy of science: J. Mittelstrass, "World Pictures: The World of the History and Philosophy of Science," in: J. R. Brown and J. Mittelstrass (eds.), *An Intimate Relation: Studies in the History and Philosophy of Science. Presented to Robert E. Butts on His 60th Birthday*, Dordrecht and Boston and London: Kluwer Academic Publishers 1989 (*Boston Studies in the Philosophy of Science*, vol. 116), pp. 319–341, particularly pp. 322–330 (2. The World of Science).

and the system of fixed stars (with a daily rotation about the axis of the heavens). The geocentrism of the *Aristotle world* is a result of the Aristotelian theory of elements or the theory of natural positions. That a heavy body falls to the earth is a result of the center of the cosmos' being the natural position for this body, i.e. the motion of heavy bodies is not toward the earth (this is only *per accidens*), but toward the centre of the cosmos (*per essentiam*).

In opposition to the atomistic conception of the constant movement of atoms, the *Aristotle world* is characterized by the notion that every movement requires a mover. Thus, not only the change of motion, but also the uniform motion of a body requires a causal force. This force must either reside in the moving body itself (in the form of a motivating "soul" or as a natural movement) or exist in direct contact with it; action at a distance is not permitted. The place of atoms in atomistic conceptions is filled by so-called *minima naturalia*, i.e. the smallest particles of matter that place a natural limit on its divisibility without altering its substantial form. Correspondingly, all matter has quantitative *minima* that possess the characteristics of macrobodies made from it. These *minima* also possess a characteristic size, though their geometric form is not predetermined. In chemical processes, *minima* in immediate proximity to each other constitute a *qualitas media*, which is the basis for the *forma mixti* of matter which possesses a particular substantival form. (According to atomistic conceptions, all that changes in chemical processes is the configuration of the smallest particles, which lack qualitative characteristics and whose geometric form is constant.)

The *Aristotle world* is thus characterized by a high degree of *experiential evidence*. The scientific propositions describing this world are confirmed by the experience acquired in everyday life, or are derived through generalizations made on the basis of experience. Examples of this are (1) the Aristotelian law of gravitation, according to which the velocity of a falling body is proportional to its weight and inversely proportional to the density of the medium, (2) the above mentioned Aristotelian "law of inertia," which states that all things moved have a mover, and (3) the Aristotelian theory of elements with its familiar concepts derived from the experience of daily life, for example, "above," "below," "natural," and "unnatural" (as in the case of violent movements that run counter to natural movements). The *Aristotle world*, moreover, is always in the process of becoming a natural order, embedded in the inner teleology of this world or the teleological nature of all things. This natural order never appears as a perfect state, but it is constantly present in the form of an astronomically ordered, supralunary world. In other words, disorder as well as the tendency to order is the normal state of the (sublunary) world. It is the world of experience and hence – despite physics and natural philosophy which seek to interpret it – a very human world.

As opposed to the *Aristotle world*, a hermetic world – by which is meant the world of alchemy, astrology, and parts of natural philosophy in the Renaissance – is a world of mysterious interactions. Occult powers and living substances take the place of the simple bodies characteristic of the *Aristotle world*. Nature consists of different combinations of primary substances that originated in undifferentiated primordial matter. At the same time, these combinations are conceived of as developmental processes that man can accelerate or retard, though always with methods that "imitate nature," for example, by "refining" metals and other substances (*transmutatio*). Inorganic processes are viewed analogously to organic processes. Explanations of the world take the shape of allegorical interpretations: coming into being and passing away as birth and death, separation and unity as the polarity of the sexes (the *conjunctio* as sexual union or the hermaphrodite as the overcoming of sexual differences).[2]

This conception finds its cosmological expression in the correspondence between *macrocosm* and *microcosm* which interprets the world in antiquity and in the hermetic tradition as a great organism mirrored in the microcosm, particularly in man: "what is below is like what is above; what is above is like what is below: both reveal the miracle of the one."[3] The influence of the macrocosm on the microcosm corresponds to the everpresent assumption in magical thought that it is possible to effect a change in the macrocosm through changes in the microcosm. This conception, as the "sympathetic" relationship between all of the parts of the world, is still at work within the context of natural philosophy in the Romantic period: man as a microcosm "in which the universe looks at itself."[4]

In a *Hermes world* everything becomes a riddle or a key to solving its secrets. The familiarity of the *Aristotle world* gives way to a demonic world that is only accessible through ritual and mystical forms of knowledge. The scientist becomes in this way the mediator between two worlds, a life-world and a hermetic world, and at the same time the real "addressee" of his own hermetic knowledge. The alchemical *separatio* reproduces itself as the *separatio* of the material and mystical body (the "diamond body") in the scientist. It constitutes the actual *magisterium*, i.e. the "great work," the self-development or spiritualization of man. Thus the hermetic world stands not only in opposition to the familiarity

2 See Chr. Thiel, "Alchemie," in: J. Mittelstrass (ed.), *Enzyklopaedie Philosophie und Wissenschaftstheorie*, 2nd edition, vol. I, Stuttgart and Weimar: J. B. Metzler 2005, pp. 75–83.

3 The first sentence of an apocryphal text attributed to Hermes Trismegistos. See Chr. Thiel, "Makrokosmos," in: J. Mittelstrass (ed.), *Enzyklopaedie Philosophie und Wissenschaftstheorie*, 2nd edition, vol. V, Stuttgart and Weimar: J. B. Metzler 2013, pp. 186–189.

4 J. J. Wagner, *System der Idealphilosophie*, Leipzig: Breitkopf und Haertel 1804, p. LIII.

of the *Aristotle world*, but also in opposition to the mechanistic world that in the modern age begins to supplant both the *Aristotle world*, as well as Aristotelian physics.

The foundation for this mechanistic world picture is Isaac Newton's world. In this world it is only (gravitational) mass that moves in absolute time, through absolute space. Matter and space are the real elements of this world. The smallest particles of matter, hence the actual atoms, combine to build complex formations or second degree particles. Several of these combine in turn to become third degree particles and so forth. The inner structure of matter is thus characterized by a complex hierarchy of particle formations. These formations are not massive corpuscles, but contain empty space. As the order of the particle hierarchy expands, the amount of empty space in them increases while the extent of solid matter decreases correspondingly. Matter in the world is thus only seemingly solid. In fact, the world is a vacuum for the most part. The actual amount of solid matter in the universe could fit into a nutshell (atomistic nutshell theory).

Characteristic of the *Newton world*, moreover, is the assumption that a fundamental dualism exists between passive matter and active immaterial principles. According to this notion, which can be traced back to Cambridge platonism and hence to hermetic conceptions of the world, matter can be the origin only of mechanical effects, that is, effects mediated by pressure and impulse. Matter itself does not exert force, but only withstands the effects of forces (through its own inertia). Gravitational pull, in particular, is not a trait of matter. Gravitation has more the status of an active principle and finds its origin in a non-material ether that exerts an effect on matter. Matter, "inanimate and brute," is not able to guarantee even halfway stable processes of development through its essential characteristics. Since in this world a general principle for the conservation of energy does not hold, mechanical interactions lead to a steady loss of movement, which cannot be fully compensated for by the active principles that bring forth new movement. All the regularly functioning causes (material or immaterial) taken together would not be able to impede the movement of the world toward disorder and chaos. The stability of the world, i.e. compensation for the energy loss, is a matter only for God or an occasional divine intervention in this world.

The nutshell theory of matter on which this world is based corresponds, as regards its concept of space, to a container or arena theory. The space of the *Newton world* is not formed by spatial relations of material bodies (concept of relational space), but exists "in and of itself" as an ontological entity on the same level as matter. Space is independent of matter. In proving the existence of inertial forces, Newton attempts to endow the related concept of absolute space, i.e. the conception of a stationary system of coordinates that differs from the mere relative state of rest between bodies, with experimental content. He himself tried to show that the centrifugal forces generated by rotational movement cannot be traced back to relative rotations of whatever type, i.e., they have to be conceived of as "true" rotations, as rotations against an absolutely stationary space. This absolute space is analogous to the *sensorium Dei*, i.e. the omnipresence of God put into law (following here the Cambridge platonists as well). Just as the mind of man can receive sense impressions through its presence in his brain, so God perceives the processes in the universe through his presence in absolute space.

The "mechanism" of the *Newton world*, expressed in a mechanics of gravitational movement, in Newtonianism not only determines how inorganic nature is understood but also proliferates itself in the organic, psychic, and social cosmos. In the theological aspects it still retains, this mechanism documents their fundamental dispensability. The criticism of the effects of occult powers (qualities) in a hermetic world also applies to Newton's theological legitimations. The *Newton world*, the quintessential "mechanization of the world picture," becomes a "world of machines" – with God as a retired engineer.[5]

In contrast to the concept of absolute space in the *Newton world*, a concept of relational space is dominant in Einstein's world. Here space is constituted only by matter, with energy also being matter. In order to do justice to the special effects of rotation discovered by Newton, Albert Einstein refers to *Mach's principle*, which considers the centrifugal forces not as the result of true rotation (rotation against absolute space) but as the effect of rotation relative to distant masses (that is, the center of gravity in the universe). Einstein's *general theory of relativity* attempts to give Mach's programmatic idea a physical dimension in order to establish the validity of a theory of relational space in terms of epistemology as well as physics.

Particularly relevant philosophically is the idea of a *geometricization of nature*. In the *general theory of relativity*, gravitation (with certain restrictions) is no longer conceived of as a force that diverts bodies from their natural trajectory, but as an entity that is inseparably bound up with the structure of space and time. If one examines the trajectory of a body in the field of gravitation from an adequate standpoint, one would recognize that this body actually follows the most linear trajectory. Later Einstein also tried to apply this idea to electromagnetic forces in order to achieve a unified theory of gravitation and electrodynamics. The central idea was that all interactions between particles can be traced back to space-time structure. This means that a particle has an effect on space-

5 E. J. Dijksterhuis, *De Mechanisering van het Wereldbeeld*, Amsterdam: Meulenhoff 1950, p. 539 (*The Mechanization of the World Picture*, Oxford: Oxford University Press 1961, p. 491).

time, which in turn has an effect on a second particle and in this way mediates the interaction between the two.

Matter itself, however, is not included in the unification. Particles are singularities of the fields, i. e., particles are not themselves solutions to the field equations. Where particles are, the field equations are not valid. Both ponderable matter as well as fields can be viewed as matter in a broader sense. Both possess physical reality in the same sense. Since the field of gravitation in particular is of this type and at the same time is identified through the *general theory of relativity* with the space-time structure, matter is everywhere where space is. The *Einstein world* is not virtually empty like the *Newton world*, but full like a *Cartesian world*. Like the *Newton world*, however, it is subject to deterministic considerations in the form of a unified theory of interactions. Thus, in the *Einstein world* there are no essentially accidental elements; everything is predetermined from the beginning and takes place necessarily. God does not throw dice.

It is characteristic of the *special* as well as the *general theory of relativity* that the essential geometric quantity is a four-dimensional metric interval. This can be divided into a spatial and a temporal component, and yet this division is dependent on the system of coordinates used. Einstein draws the conclusion from this situation that the "transient now" (the idea of a shifting present) possesses no objective meaning. He draws the same conclusion from the symmetry of equations in mechanics and quantum mechanics against time reversal. In all elementary processes there is no difference between past and future. Such a difference is a mere illusion. In reality there is no development, no actual change. All that is real is a static, four-dimensional state of being. In this sense the *Einstein world* is neither Aristotelian nor hermetic nor Newtonian, but Parmenidean.

The examples cited here demonstrate the power of science to constitute worlds and generate world pictures. There can be no doubt that it fulfills this role. But the brief discussion of the *Plato*, the *Aristotle*, the *Hermes*, the *Newton*, and the *Einstein world* demonstrate even more. In the final analysis, these examples reveal not only that science makes worlds, but there is also a certain relativity of scientific worlds. Each of these worlds, with respect to the scientific (or philosophical) view of things on which they are based, has its own plausibility, and each is somehow consistent. What they show – quite apart from the fact that nobody today, in a scientific world as well as in the life world, wants to go back to an Aristotelian or an hermetic world – is a growing loss of perceptibility and an increasing distance to what is understood by nature. There is no room left for perceptibility in an *Einstein world*, let alone the world of quantum mechanics, a *Planck* or *Heisenberg world*, in which particles no longer move on trajectories like in classical physics and the law of causality no longer holds. The same is true with the concept of nature. Newton's concept of nature is different from that of Aristotle, and Einstein's concept of nature (if there is one) is different from that of Newton. Physics, as it appears, has lost the concept of nature, it is biology which may bring it back.

9 Causality in Greek Thought

Our philosophical as well as our scientific thought has its beginning in Greek thought. This is also true with respect to the concept of causality. It is a central concept of philosophy as well as of the sciences, especially the natural sciences. And as the Greeks saw in natural philosophy an essential part of their thought, they will certainly have had some ideas about causality. How, indeed, should one answer why-questions, around which early rational thought begins to revolve, without hitting on the relationships between reason and consequence or cause and effect, or without drawing these distinctions?

Logic deals with reason and consequence, natural philosophy with cause and effect, for instance by distinguishing between a *principle of causation* (*nihil fit sine causa* / nothing happens without a cause) and a *law of causation* (same causes have same effects). Are these distinctions Greek distinctions? For Greek logic, in particular in its Aristotelian form, which deals with the relation between reason and consequence, the answer is Yes. But is this also true for Greek natural philosophy, which deals with cause and effect? Things are much less clear here, especially if we look at it from the perspective of modern conceptions of causality.

Indeed, to trace the concept of causality in Greek thought means to go strange ways. The Greek conception of causality has little to do with modern conceptions of causality; Kurt von Fritz, among others, pointed this out already in 1961.[1] In fact, the questions that get answered with considerations of causality are often different from those asked later. At first sight, that may appear surprising, as, again, natural philosophy makes up a large part of Greek philosophy, and the analysis of the relation between cause and effect constitutes the heart of explanation of natural processes. At any rate, this is how things have presented themselves since Galileo, that is, since the emergence of modern natural science. Yet they do not, even in their beginnings, have to be *part* of Greek thought, let alone the yardstick against which this thinking has to be measured.

This will become clear in what follows. Let us take up the relevant considerations in Plato and Aristotle, including perspectives from general philosophy of science. This will not be a description of all that might fall under the concept of causality in Greek thought, but will concentrate on what is exemplary in this thought in the context of this concept.

1 K. von Fritz, "Der Beginn universalwissenschaftlicher Bestrebungen und der Primat der Griechen II," *Studium Generale* 14 (1961), pp. 601–636, here pp. 622ff..

https://doi.org/10.1515/9783110596687-009

9.1 Becoming and passing away

Even if the concept of causality may not be central in Greek philosophy, this does not mean that causal relationships, in particular natural ones, did not receive any attention at all. Already Presocratic philosophy, dealing with natural philosophy, is interested in *forces* and *effects*, for instance, by looking at how things have effects in virtue of their properties (effects on themselves or on others). Furthermore, the physical events, including in particular cosmic events, are mostly perceived in light of regularities, which have a basis in experience, on the one hand, and in (mathematical) models on the other, which in turn privilege kinematic, non-causal explanations. With respect to the effects of properties, Anaxagoras says, for instance: "water gets released from the clouds, earth from water, stones are solidly composed of earth, under the influence of the cold."[2] And Empedocles says: "When water and earth and air and sun mixed with each other, the forms and types of mortal things were generated (...) by being conjoined by love, precisely as many as there are in the world today."[3]

Among the regularities dominating all events is the following, according to Heraclitus: "Everything is exchangeable against fire and fire against everything, just as goods against gold, and gold against goods."[4] Similarly, Empedocles writes: "At some time, the whole comes together in love to be one – limbs which the body possesses in the most vivacious stage of life; the other time, by contrast, torn apart by bad feuds, everything gets driven apart, as life collapses. It is the same with the bushes and the fish dwelling in the water, and the animals lairing in the mountains and the hovering birds."[5] Here it is love and war, or disagreement, which constrain and determine all regularities. Some of it indeed sounds like the formulation of a principle of causation in general, for instance when Leucippus says: "Nothing comes into existence at haphazard, but all as a result of a *logos* (λόγος) and by necessity."[6]

But again, it is considerations of a cosmic order that are of primary interest, and not those of an analysed causality. That events occur in nature, in the cosmos, following certain patterns, is more important for the understanding of cause and effect than for its analysis. Perhaps one could say that in these first attempts at explanation in natural philosophy, the step from description (of

2 59 B 16 (The Presocratics, quoted according to: H. Diels, *Die Fragmente der Vorsokratiker: Griechisch und Deutsch*, 6th edition, Berlin: Weidemann 1951–1952).
3 31 B 71.
4 22 B 90.
5 31 B 20.
6 67 B 2.

what is the case) to explanation (of why something is the case) occurs too early; explanation oversteps itself. In other words, this nascent rationality, in a certain sense, still follows mythical thinking, which in its own way just skips this step – even in the assumption of hidden intentionalities that take on the role of divine action, for instance in the form of love and war (Empedocles) or an effective reason (Anaxagoras) as supposed causalities.

Such ideas remain an element of developed Greek thought. It is almost reminiscent of Presocratic conceptions of harmony when Aristotle explains, in the first book of the *Physics* that "what is in tune must come from what is not in tune, and *vice versa*; the tuned passes into untunedness – and not into any untunedness, but into the corresponding opposite."[7] Becoming and passing away, as the Presocratics had said, are the basic processes in nature, the explanation of which the search for material and immaterial principles that become effective in becoming and passing away is aimed at. This is still far from a precise conception of causality; in the framework of ideas of regularities inherent to Greek thought, this would mean for instance that everything is governed by the conditions that the cause has to precede the effect in time and that there is an empirical connection between cause and effect.

9.2 Platonic causes

Plato's views, developed in the context of the theory of ideas and presenting themselves as a conscious break with the prevailing natural philosophy, are often considered an important step in this direction. The framework is given by the well-known intellectual autobiography of Socrates in the *Phaedo.*[8] Socrates had dedicated himself to the search for the causes of becoming and passing away in his youth, considering problems such as whether animals originate by the cold and the warm turning to dirt, or whether we think with our blood or with air or with fire, or with none of them, but our brain instead.[9] That is good Presocratic thinking. After all, well-confused by all of this, Socrates heard Anaxagoras saying that reason (νοῦς) is the cause of everything.[10] Cause and reason – these fit together well following Socrates, in particular since causal explanations, which a philosophy now beyond the mythical is

7 *Phys.* A5.188b12–15. Translation by R. P. Hardie and R. K. Gaye, as found here: http://classics.mit.edu/Aristotle/physics.1.i.html

8 *Phaedo* 95e-101e.

9 *Phaedo* 96b.

10 *Phaedo* 97b/c.

searching for, themselves claim to be rational explanations. Indeed, reason itself now enters the scene as the general cause, as Socrates believes himself to discern, or as a *teleological principle* that dominates all things, including natural processes. But Socrates will be disappointed. Reason, according to Anaxagoras, only gets the things moving at the beginning; after that, everything goes the way that Anaxagoras's predecessors had already taken in their attempts to explain becoming and passing away.

In this situation, which would only mean a futile search for empirical causes, that is, empirical causes in the empirical world, Socrates, as he says, takes refuge with the *logoi*, that is, in the constructions of reason.[11] In them he believes now that he perceives the "truth of things" (τῶν ὄντων τὴν ἀλήθειαν).[12] Socrates, in other words, has ceased merely looking at things, that is, looking for explanations in the empirical realm.[13] His search for the truth of things is now a search for the *natures* of things in themselves and no longer a search for causal connections that explain empirical processes. These natures may not be empirically seen. The empirical facts are, to use the language of the allegory of the cave, mere images on the wall; reality happens in thought. Even if the concept of cause (τῆς αἰτίας τὸ εἶδος) is still being employed,[14] it occurs no longer in the sense of an explanation of causal processes, but as an explication of the relation between things and corresponding ideas, that is, it is about the "participation" (μέθεξίς) of things in their ideas.[15] These relations are conceptual, not empirical.

The result is that Platonic philosophy moves away from an explanation of causalities and toward a theoretical form of explanation. The theoretical now occupies all of philosophy's attention: the empirical has to wait – for Aristotle.

Nonetheless, interpreters of Plato have tried again and again to read Plato's turnaround in the *Phaedo* as introducing a new concept of causality, which also takes account of the empirical, that is, the explanation of empirical processes. For instance, according to Gregory Vlastos,[16] Plato's new conception consists in the fact that empirical phenomena may be explained as "logically necessary" in the same way that ideas are; that is, an empirical effect follows an empirical cause with the same necessity with which 2 + 2 gives the result 4 (numbers being

11 *Phaedo* 99e.

12 *Ibid.* 99e6.

13 *Phaedo* 99d.

14 *Phaedo* 100b.

15 *Phaedo* 100c.

16 G. Vlastos, "Reasons and Causes in the *Phaedo*," *The Philosophical Review* 78 (1969), pp. 291–325. Also in: G. Vlastos, *Platonic Studies*, Princeton N.J.: Princeton University Press 1973, pp. 76–110.

conceived as ideas in line with the theory of ideas of numbers). But this was hardly the real interest of Plato (in the sense that it was indeed Leibniz's philosophical interest to reduce truths of matters of fact to truths of reason). Plato is no longer interested in the explanation of (empirical) effects by tying them to their (empirical) causes. For, as we have just seen, he is interested in the relation between things and their ideas, on analogy with a relation between an original model and its representation. Representation assigns to an idea (as a universal original) an object (as a particular representation). Cognition is "the result of the execution of the mapping (which is assumed to exist) which goes inversely from the representation to the original."[17] This is the relation that Plato is interested in, not the one between (empirical) cause and (empirical) effect. At most, one could say the following, due to David Sedley, to keep up the causal terminology: "that it is the F [the appropriate Form] which causes F things to be F."[18] In other words, the properties of things are caused by the corresponding ideas, or, as Sean Kelsey says, "Socrates' Form-hypothesis maintains that things participate in Forms as the result of causes whose object is to make them do that"[19]. No doubt causality is being used in a metaphorical sense here.

For Plato, the interest shifts from the empirical (with its properties and regularities) to the conceptual, and this move is explicitly justified by the insufficiency of the then prevailing attempts at the explanation of causality. That the things are as they are, including their effects on themselves and other things, is not due to empirically given properties and regularities, but due to the "participation" of the things in their ideas, which are, to use causal terminology once again, the causes of the given properties and regularities. A conscious break is made with the prevailing natural philosophy, to the extent that this may be understood, in its Presocratic form, as an investigation of causality. And when Plato surprisingly takes up the topic of research into nature again, as in the *Timaeus*, this does not happen in the sense of a revision of this conception, but, on the contrary, as its consistent application or extension.

In the *Timaeus* nature is not described as given, but as the work of a demiurge who has created it following an ideal plan or pattern, the "cosmos" of the Platonic ideas. Following the model of a "perfect being," the cosmos emerges as

---

17 K. Lorenz, "Abbildtheorie," in: J. Mittelstrass (ed.), *Enzyklopaedie Philosophie und Wissenschaftstheorie*, 2nd edition, vol. I, Stuttgart and Weimar: J. B. Metzler 2005, pp. 6–8.

18 D. Sedley, "Platonic Causes," *Phronesis* 43 (1998), pp. 114–132, quoted is p. 127.

19 S. Kelsey, "Causation in the *Phaedo*," *Pacific Philosophical Quarterly* 85 (2004), pp. 21–43, quoted is p. 23.

a being with soul and reason,[20] as a visible God in the form of a perfect sphere.[21] His soul, the "world soul," has an astronomical being; it is formed by the (mathematical) order of the planetary trajectories. At the same time, the celestial bodies figure as "tools of time,"[22] the time (χρόνος), having come into existence together with the heaven, is a representation of eternity (αἰών).[23] The celestial bodies are "visible and emerged Gods,"[24] the Earth the "first and most honourable Goddess within the heaven."[25] Man, in this cosmos, is a "plant" which "has its roots not in the Earth, but in the heaven;"[26] he connects the Earth with the heaven (related to him).[27] And so on.

What looks like a mythical relapse is in reality the consistent application of the theory of ideas to cosmology. In its classic conception, the theory of ideas does not allow for a science of nature, that is, of the world of becoming and passing away. Here, however, in the *Timaeus*, the cosmos is presented as the outcome of a planned process of creation, as an *artifact*. And artifacts allow, in the Platonic conception, an explanatory recourse to relations within the theory of ideas. That is, the thought that the physical world may be amenable to explanation by perceiving it as a world "put into action," as an artifact, draws the cosmological language of the *Timaeus* nearer to the language about geometrical ideas and their (always imperfect) realizations, which form the standard introductory text on the theory of ideas. In the use of models, for instance mechanical models of planetary motion, but also stereometric models of the formation of elements, the cosmological conception of Plato gains, at least in principle, a methodological profile that, without this connection, is at risk of foundering as pure fancy. This is research into nature not in the sense of research into causality, but as a construction of a world as it corresponds to ideal circumstances. Some have therefore suggested that the demiurge is a metaphor for Timaeus, the "author" of the lecture, and that he in turn represents the (constructing) mind or the (con-

20 *Tim.* 30a/b. For the following, compare: J. Mittelstrass, "Die Kosmologie der Griechen," in: J. Audretsch and K. Mainzer (eds.), *Vom Anfang der Welt: Wissenschaft, Philosophie, Religion, Mythos*, Munich: Beck 1989, pp. 40–65, 208–210, also in: J. Mittelstrass, *Die griechische Denkform: Von der Entstehung der Philosophie aus dem Geiste der Geometrie*, Berlin and Boston: Walter de Gruyter 2014, pp. 43–71.
21 *Tim.* 34a; cf. 68e, 92c.
22 *Tim.* 42d.
23 *Tim.* 38c.
24 *Tim.* 40d.
25 *Tim.* 40c.
26 *Tim.* 90a.
27 *Ibid.*

structing) reason.[28] The world – so the result of the corresponding analyses goes – is a work of the explaining mind, or the explaining reason, or, to put it differently: the world is *created in theory.*

Platonic *theoria* is, in its systematic conception, science at the frontiers of constructive and thus, in this sense, nonempirical knowledge formation. At the same time, the Greek *constructive model* of rationality has its foundation in the Platonic identification of this form of knowledge with the concept of scientific rationality.

9.3 Aristotelian causes

The Greek history of causality does not end with Plato and his transformation of empirical research into conceptual research. Aristotle soon reversed this transformation, not by going back to the Presocratic ideas of becoming and passing away, but by replacing Plato's theory of the nature of things with a theory of the *perspectives* under which we see and explain things.

The point of departure is the concept of substance.[29] According to Aristotle, substances are initially something made up of matter and shape, or a concrete "this-there" (τόδε τι). They are not displaced natures of things but the things themselves, as he emphasizes in contrast to the Platonic position.[30] Substances – Plato's ideas – return into the things. And for Aristotle, it is primarily the question about the nature of things that determines inquiry, and not the demand for explanation of causal relations.

The return occurs in two ways: in the context of the so-called Substance-Accident scheme, and in the context of logical analysis. The Substance-Accident scheme says that substance is defined as bearer of properties ("a rose is red") and as bearer of appearances ("the trajectory of Mars is irregular"). In both cases, the knowledge of things is supposed to go beyond the analysis of arbitrary (accidental) features to essential (substantial) features. This distinction, in turn, has a logical as well as an ontological status, depending on whether one focuses on distinctions such as those between features that do belong to the definition of an object and those that do not, or whether one interprets the possession of fea-

28 K. J. Lee, *Platons Raumbegriff: Studien zur Metaphysik und Naturphilosophie im "Timaios,"* Wuerzburg: Koenigshausen & Neumann 2001, pp. 47 ff..

29 For the following, compare: J. Mittelstrass, "Die Aristotelische Metaphysik," in: R. Brandt and Th. Sturm (eds.), *Klassische Werke der Philosophie: Von Aristoteles bis Habermas*, Leipzig: Reclam 2002, pp. 14–37.

30 *Met.* A9.991b2–3.

tures realistically, in the sense of properties, as an inclusion relation (that of a property being included in a substance).

The definition of *ousia*, substance, in the strict logical sense proceeds via the analysis of specific subject terms, which may not themselves occur as predicate terms:[31] *Ousia* is "what is not asserted of a substratum, but that of which everything else is asserted"[32] ("substratum" [ὑποκείμενον] is here used in the sense of logical subject). This again allows the interpretation of *ousia* as the capacity of being a property-bearer ("Socrates is wise" is read as "Socrates is the bearer of the property of wisdom"). This double meaning remains part of the tradition of metaphysics with Boethius and his distinction between the first and second substance.[33] According to him, the *first ousia* is the object itself, with its accidental features; this would be the primary meaning of *ousia*. The *second ousia* is the concept that defines the object (its "essence"). It is well known that for this concept, Aristotle uses the formula: "what it means, to be this" (τὸ τί ἦν εἶναι) which is notoriously hard to translate; but it is the *characteristic concept* of an object that is intended, that for which something (on a conceptual level) is what it is.

Further determinations of the concept of substance occur with the pair of concepts *dynamis* ("potentiality") and *energeia* ("actuality"), which are used to express how things become what they are. *Dynamis* is understood, on the one hand, as the capacity to bring about changes (in another object), and on the other hand, as the capacity to be the passive subject of changes;[34] this is the standard version of an Aristotelian causality. *Energeia* is understood as the actualized determination, while actuality means the attained telos of an object: "Aim (τέλος) is the form (μορφή), completed is, what has reached its aim."[35]

The connection of the Aristotelian concept of substance with the concept of cause, finally, is effected using the concept of *arche* (ἀρχή). The *arche*, "the first out of which anything develops,"[36] is also the cause, the reason, why something is the way it has come to be, that "without which the subsequent may not be."[37] According to Aristotle, this cause gets too hastily identified by the Presocratics with the matter out of which things are made, or with the ideas as the displaced natures of things, by Plato.

---

31 *Cat.* 5.2a11–13.
32 *Met.* Z3.1029a8–9.
33 Cf. *Cat.* 5.2a11–19.
34 *Met.* Δ12.1020a2–3.
35 Met. Δ24.1023a34.
36 Cf. *Met.* Δ1.1012b34 ff..
37 *Met.* Δ2.1013a24 ff..

Thus it is surprising that, as is well known, Aristotle seems to know several concepts of cause, normally grouped under the title "doctrine of the four causes."[38] According to this, cause is "that out of which something develops,"[39] "the form (εἶδος) or the original (παράδειγμα),"[40] "the beginning (ἀρχή) of a change (μεταβολή)"[41] and "the aim (τέλος) or the what for (τὸ οὗ ἕνεκα)."[42] According to modern or even to early modern opinion, only the third meaning ("the beginning of a change") would represent a cause in the narrow sense. Indeed, the modern criticisms are mostly addressed against the use of the concept of *telos*. But what appears from the modern perspective to be a turn towards a more scientific approach in fact represents, from the Aristotelian perspective, an impoverishment of understanding and inquiry. And this derives from Aristotle's way of conceiving all explanation and understanding on the basis of perspective.[43]

According to Aristotle, physics is primarily a *principled analysis*; that is, physics – and this is where Aristotelian causal analysis begins – introduces perspectives under which natural phenomena (processes, states of affairs, events) are perceived, described, and explained. Thus Aristotle transfers the problem of physical explanation from the object level ("what are the principles of things?") to a meta-level ("according to what perspective do we look at and explain things?"). The point of departure here is that of inquiry, in particular, of why-questions. Why-questions, or questions of the type "what is the case?" or "why is something the case?" may, according to Aristotle, be put in four different ways, the answers to which, in Aristotelian terminology, label causes or principles: "of what does something consist?," "what is something?," "how is something caused?" and "what does something serve for?" The answers point to the concepts of matter (*causa materialis*), of form (*causa formalis*), of efficient cause (*causa efficiens*), and the aim or purpose (*causa finalis*). According to Aristotle, an analysis, in the form of an inquiry into things and states of affairs, is incomplete, just as are their explanation and the grasping of them, unless all these questions are answered. However, the completeness of this set of questions is not claimed.

---

38 *Phys.* B3.194b23 ff.; *Met.* A3.983a24–32.
39 *Phys.* B3.194b24–25.
40 *Phys.* B3.194b26–29.
41 *Phys.* B3.194b29–31.
42 *Phys.* B3.194b32–33.
43 W. Wieland in particular has elaborated on this (*Die aristotelische Physik: Untersuchungen ueber die Grundlegung der Naturwissenschaft und die sprachlichen Bedingungen der Prinzipienforschung bei Aristoteles*, Goettingen: Vandenhoeck & Ruprecht 1962, 2nd edition 1970).

In this connection, Aristotle reminds us, as he does often, of ordinary language, in which different meanings of cause (αἴτιον) are observable.[44] Thus the project is one of the analysis of meaning, not of causality in the direct sense. The why-question (διά τι), so Aristotle says, may be stated in four different ways and, accordingly, may be answered in four different ways. The unity of the meanings that are materialized in this form is not given in one overarching principle, that is, *the* meaning of cause, but in different functions that the question about the why may assume.[45]

Traditionally, however, the concept of *telos*, that is, the reply to the question about the what-for, has been conceived as such an overarching principle (for instance, by St. Thomas Aquinas in his commentary on the *Physics*[46]). Aristotle is not completely innocent of this misunderstanding, for instance, when he says, that the what-for always wants, as cause, to be the aim of the other causes.[47] Appropriately enough, Wolfgang Wieland observes that "from the fact that the Telos always requires the other causes, (....) it does not follow that it is itself a cause in a higher sense. It is not the other causes which are meant by the 'other' (τἀγαθὸν τῶν ἄλλων), but, in a more general sense, all that logically precedes the purpose and which it requires, primarily, thus, the means."[48] Again, Aristotle turns against a restriction on the principle of matter. Through such a restriction purposefulness would not be achieved and everything, even the purposeful, would be left to chance.

This explanation makes clear that in the Aristotelian analysis of the concept of cause, the concept of effect does not play any role in the sense familiar to us, nor in the sense considered to be constitutive in later discussions of causality. It does not figure explicitly as a concept complementary to that of cause, for it does so at most in the determination of that *for which* the causes are causes.[49] This in turn is also due to the fact that Aristotle, just as Plato, is mostly interested in the causes of things, not in the causes of processes, that is, of caused happenings. The Aristotelian concept of cause appears strange to us not because it is so closely linked with the concept of purpose (the concept of *telos*), but because the complementarity of

44 *Phys.* B3.195a29.

45 Cf. W. Wieland, *Die aristotelische Physik*, pp. 261 ff..

46 *In octo libros de physico auditu sive Physicorum Aristotelis commentaria, Opera omnia. Editio Leonina II* (1884), ed. F. Angeli and M. Pirotta, Naples: M. D'Auria 1953, p. 5,11.

47 *Phys.* B3.195a23–25.

48 W. Wieland, *Die aristotelische Physik*, p. 263.

49 *Phys.* B3.195b7.

cause and effect plays only a subordinate role.[50] But this may be explained directly in light of the Aristotelian analysis of causation itself. Aristotle stated verbatim: "Our inquiry concerns the discovery, and for the discovery of a thing we require the realization of its cause, and that means: of the first cause."[51] Besides, Aristotle remarks that even Plato has labelled the what-for – that is, the teleological principle – as cause (αἴτιον), but not in the context of a principled analysis, as he does[52], nor in the sense of a reconstruction of the structures of experience, to which, in the Aristotelian doctrine of causes, the principles addressed here refer, in the form of inquiring questions.

In a more general epistemological sense, we find, in opposition to Plato's construction model, the Platonic *theoria*, a *reconstruction model*, the basis of which is the Aristotelian *empeiria*. This plays a constitutive role in scientific knowledge formation, to the extent that, according to Aristotle, all scientific knowledge is based on pre-scientific, pre-theoretical knowledge. The central concept is that of *phenomenal* experience, as opposed to the concept of *instrumental* experience, in which, for instance in physics since Galileo, experiences are not given but produced experimentally. It is, indeed, this fact – the existence of different concepts of experience and their methodological use – that constitutes the real difference between the Aristotelian and modern physics. It is not the supposed fact that Aristotelian physics is not an empirical physics. Aristotelian physics is actually as empirical as modern physics, but in a different way. The controlling concept is not experimental or instrumental but phenomenal experience.

That too is evident in almost an exemplary manner in the Aristotelian concept of cause, as opposed to both the Platonic and the modern concepts of cause. According to Aristotle, the world of phenomena already has a conceptual structure.[53] The reconstruction of this conceptual structure takes place on analogy with a principled analysis of the concept of cause dealing with the given stock of experience.

To summarize, the concept of causality is always present in Greek thought, but it does not find terminological or conceptual clarity. In Presocratic thought, it shows itself in the search for properties or qualities and regularities that are sup-

---

50 Wieland (*Die aristotelische Physik*, p. 266) points out that the concept of αἰτιατόν, which could correspond to the concept of effect, only appears in two peripheral passages (*Met.* K8.1065a11, *an. post.* 16.98a36, b3), none of which is in the *Physics*.

51 *Phys.* B3.194b17–20.

52 *Met.* A7.988b6–8.

53 See G. E. L. Owen, τιθέναι τὰ φαινόμενα, in: *Aristote et les problèmes de méthode*, Louvain: Publications universitaires and Paris: Béatrice-Nauwelaerts 1961, pp. 83–103.

posed to explain becoming and passing away. In Platonic thought, it changes fronts from the "physiologists," who were looking for materialistic causes in particular, to the "friends of the idea," for whom everything empirical seems to dissolve into the conceptual and all knowledge dissolves into theory. In Aristotelian thought, talk of causes regains its empirical meaning, but not in the sense of the later concept of causation (the concept of effect in the context of cause and effect remains inefficacious), but in taking the form of a principled analysis that operates between things and their theory.

10 Scientific Truth, Copernicus, and the Case of an Unwelcome Preface

Science is the expression of universal claims to validity, and this is true both in the sense that it is a special form of knowledge formation, that is to say of the scientific production of knowledge, and in the sense that there is a specifically scientific ethos, which is also the moral form of science. The orientation towards truth typical of the former follows the orientation towards truthfulness of the latter. This is to say, quite simply, that *truth* determines the scientific form of knowledge, whereas *truthfulness* determines the moral form of science, which as a result belongs to the form of life of the scientist, to his ethos.

Where the notion of scientific truth leads into philosophy of science, becoming more and more subject to a relativistic view – although it should, as a guiding *idea*, play an essential role – the notion of a scientific ethos leads to ethics: in this case to the ethics of science. This ethics deals (1) with research-focused ethical problems and principles – for example, problems arising within stem cell research and reproductive medicine; (2) with application-focused ethical problems and principles – for example research in nuclear physics which leads to products whose manufacture and use creates serious ethical problems; and (3) with ethos-focused ethical problems and principles since, as in other areas of human practice, falsehood and deceit also have their place in science. It is the third type of problem which will be discussed here. Both in the history of science and today, there are many examples – recall the scandal that took place in stem cell research in a South Korean institute in 2005, or the faking of research results by a German physicist at the Bell Laboratories in 2002 – of methods and results being manipulated, or plagiarized, and of publications being tampered with. If there exists an ethos of the scientist and the scientific system, it is sometimes corrupted.

With that, everything seems to have been said about falsehood and deceit in science. Wherever they occur, both the ethos of the scientist and scientific truth are violated. But is that really all? Could there not be something like the "cunning of reason" which, on devious paths, serves the scientific truth, as a science-promoting deception? Let us consider an example from the history of science.

In 1543 Nicholas Copernicus' crucial treatise, *De revolutionibus orbium coelestium libri VI*, was published in Nuremberg. The heliocentric system replaced the geocentric system, that is, the system in which the centre of the world is also the centre of the earth, which had been valid till then. Copernicus had previously tried to mathematically represent the eccentric and compensatory move-

https://doi.org/10.1515/9783110596687-010

ments of the planets in the (geocentric) Ptolemaic system using two uniformly rotating epicycles (*De hypothesibus motuum coelestium commentariolus*, c. 1510). When this did not work, he proceeded to develop a model in which some of the irregularities of the planetary motions could be explained as effects of the motion of the earth. He replaced the first epicycle and its concentric deferent with a kinematically equivalent eccentric deferent, which, however, also moves the sun from the centre of the system to an eccentric position. It is from this moment onwards that heliocentrism becomes the starting point of all further astronomic developments.

Copernican astronomy is thus the classic example of a scientific revolution – and not just one we see with hindsight, and so in a sense "discovered" by historians of science, but also one of which Copernicus' contemporaries were conscious. Galileo Galilei used it to agitate against the Aristotelian system of natural science, and Immanuel Kant likens his radical conversion of epistemological perspectives to a "Copernican revolution."[1] Indeed, even Copernicus' own assessment corresponds to this view. For him, the heliocentric model in the way he has presented it is the true model of the world. For instance, the Wittenberg mathematician Georg Joachim Rheticus, who had stayed in Frombork from spring 1539 until the autumn of 1541 to learn about the Copernican System, emphasizes in his *Narratio prima* (1540, 2nd edition 1541)[2], which contains the first short description of the Copernican system in print (even before *De revolutionibus*) that Copernicus has restored the "astronomical truth" and set out the "true system" of the world (*systema mundi*).[3] And Copernicus himself underlines this claim with his remark that he has shown the "true form of the world" (*forma mundi*).[4]

In stark contrast to this, and to the customary classification of Copernican astronomy as a scientific revolution, stands the preface to the 1543 Nuremberg edition, written not by Copernicus himself but by Andreas Osiander, a Lutheran theologian. He had prepared the first edition after Rheticus had passed the duty

1 *Critique of Pure Reason* B XVI.

2 Reprinted in: *Johannes Kepler. Gesammelte Werke*, ed. W. v. Dyck and M. Caspar and F. Hammer, Munich: C. H. Beck 1937 ff., vol. I, pp. 81–126. The second edition of *De revolutionibus* (Basel: Heinrich Petri 1566) contains the *Narratio prima*; a *Narratio secunda* had become redundant after the Copernican work had been published.

3 *Ibid.*, p. 97, cf. p. 101.

4 In his dedication to Pope Paul III., *Nicolaus Copernicus. Gesamtausgabe*, vol. II (*De revolutionibus libri sex*), ed. H. M. Nobis and B. Sticker, Hildesheim: Gerstenberg 1984, p. 4 (English edition: *Nicholas Copernicus. On the Revolutions*, Translation and Commentary by Edward Rosen, London: Macmillan 1978, p. XVI).

of overseeing the printing on to him in November 1542. At the time he was taking up a professorship in mathematics in Leipzig. Against the Copernican self-understanding and its propagandistic representation by Rheticus, Osiander explicitly emphasizes the *hypothetical* nature of the Copernican system.

Thus Osiander writes: "It is the duty of an astronomer to compose the history of the celestial motions through careful and expert study. Then he must conceive and devise the causes of these motions or hypotheses about them. Since he cannot in any way attain to the true causes, he will adopt whatever suppositions enable the motions to be computed correctly from the principles of geometry for the future as well as for the past. The present author has performed both these duties excellently. For these hypotheses need be neither true nor even probable. On the contrary, if they provide a calculus consistent with the observations, that alone is enough." He concludes: "Therefore alongside the ancient hypotheses, which are no more probable, let us permit these new hypotheses also to become known, especially since they are admirable as well as simple and bring with them a huge treasure of very skilful observations. So far as hypotheses are concerned, let no one expect anything certain from astronomy, which cannot furnish it, lest he accept as the truth ideas conceived for another purpose, and depart from this study a greater fool than when he entered it."[5] According to Osiander, then, Copernicus' achievement does not consist in having shown the "true form of the world" (*forma mundi*), but in having formulated an hypothesis which, like previous hypotheses, was suitable to represent the planetary system, just better, and more successfully.

Is this a betrayal of Copernicus? Is it possible to detect here complicity with the printer, Johannes Petreius, and the erstwhile abbot of the cloister of St. Giles, Friedrich Pistorius, who, after retiring from office, worked as editor at the printing-office?[6] In effect, the interests of the church might have influenced the classification of the Copernican system. But that is rather unlikely. Osiander did not work in secret, after all. In a letter dated April 20, 1541[7] he writes to Copernicus

5 *Ad lectorem de hypothesibus huius operis*, *Gesamtausgabe*, vol. II (appendix IV), p. 537 (English edition: p. XVI).

6 See E. Zinner, *Entstehung und Ausbreitung der copernicanischen Lehre*, 2nd edition, Munich: C. H. Beck 1988, pp. 253–254. See also H. Blumenberg, who ascribes a theological motive to Osiander: *Die kopernikanische Wende*, Frankfurt: Suhrkamp 1965, pp. 92–99 ("It becomes apparent that the contentious preface by Osiander must be understood as a principled objection against any rational claim to truth, and not just against the special case of the Copernican oeuvre," p. 92 [my translation]).

7 *Apologia Tychonis contra Ursum*, in: J. Kepler, *Opera omnia*, vols. I-VIII, ed. Ch. Frisch, Frankfurt and Erlangen: Heyder & Zimmer 1858–1871, vol. I, p. 246.

(whose preceding letter from July 1, 1540 is now lost[8]), that he might want to address the hypothetical character of the kinematic models in astronomy in the introduction: "For in this way you would mollify the peripatetics and theologians, whose opposition you fear."[9] On the same day, to Rheticus: "The peripatetics and theologians will be readily placated if they hear that there can be different hypotheses for the same apparent motion; that the present hypotheses are brought forward, not because they are in reality true, but because they regulate the computation of the apparent and combined motion as conveniently as may be; that it is possible for someone else to devise different hypotheses; that one man may conceive a suitable system, and another a more suitable."[10] Copernicus had been warned, but he did not respond to the warning. Although he also makes use of the term "hypothesis" in describing his system, he does not use it (as is common) to refer to hypothetical assumptions that could be falsified, but rather to refer to *principles* in a fundamental axiomatic sense.[11]

Deceit of the reader? Perhaps, at first sight; the authorship of Osiander's preface remains anonymous. But Osiander repeatedly refers to "the author" in his preface *Ad lectorem*, and also to the original preface of Copernicus, the *Praefatio Auctoris* which the book also contains. In turn, the author of Osiander's preface might be anonymous simply because an open intervention by a well-known Lutheran would have caused further commotion, possibly hindering the reception of the work.[12] Besides, Osiander enjoyed a scientific reputation of his own, despite the fact that he was not a scientist and had received no scientific education. Johannes Kepler, for example, referring to astronomical research, described him as "most expert on these matters."[13]

The history of science vindicates Osiander. Since antiquity (as a cosmological consequence of Aristotelian physics) a distinction had been drawn between

---

8 Mentioned in Kepler, *Apologia, loc. cit.*, pp. 245–246.

9 Translation following E. Rosen, *Three Copernican Treatises*, New York: Dover Publications 1959, p. 23.

10 J. Kepler, *Apologia, loc. cit.*, p. 246 (translation following E. Rosen, *ibid.*). The later intervention of Robert Bellarmine adressed to Galileo, who simply adopts the claims to truth of Copernicus and Rheticus in his advocacy of Copernican astronomy, needs to be understood in the same way (letter dated April 12, 1615 to the Carmelite Paolo Antonio Foscarini, *Le opere di Galileo Galilei. Edizione Nazionale*, vols. I-XX, Florenz: Barbèra 1890–1909, vol. XII, pp. 171–172).

11 See *Gesamtausgabe*, vol. II, pp. 4 (Praefatio ad Pontificem), 487 (English edition: pp. 4,7).

12 See B. Wrightsman, "Andreas Osiander's Contribution to the Copernican Achievement," in: R. S. Westman (ed.), *The Copernican Achievement*, Berkeley and Los Angeles and London: University of California Press 1975, p. 234; J. Hamel, *Nicolaus Copernicus. Leben, Werk und Wirkung*, Heidelberg and Berlin and Oxford: Spektrum 1994, pp. 230–231.

13 *Apologia, loc. cit.*, p. 246.

mathematical astronomy (which is kinematic, i.e. force-free) and physical astronomy (which is dynamic). According to Simplicius of Cilicia, a commentator on Aristotle, it is the job of physical astronomy to discover the nature of the heavens and the celestial bodies (for which Aristotelian physics provided unrivalled conditions) and the job of mathematical astronomy to prove that the planetary world really is a cosmos – that is, a system ordered according to geometric rules (which might be shown using various, even heliocentric, assumptions).[14] Indeed, arguments based on physics that one could have adduced against an Aristotelian physics, which supported the geocentric system, are missing from Copernicus' writings, which is why the Copernican system belongs, following the remarks of Osiander, to the history of mathematical astronomy, not (yet) to the history of physical astronomy.

It is only Kepler who – on the basis of a new, and truly revolutionary new approach, with which he overturns all of previous astronomy – strives for a new kind of physical argument. His formulation of a mutual attraction between two bodies, with its strength depending on their distance,[15] already points to the direction in which Galileo's kinematics is going to be extended by Isaac Newton's dynamics. Kepler's second law (the radius vector sun – planet sweeps out equal areas during equal intervals of time) is explained by Newton dynamically, by assuming a central acceleration towards the sun, the size of which may approximately be determined by Kepler's third law (the square of the orbital period of a planet is directly proportional to the cube of the semi-major axis of its orbit). Speaking purely kinematically, the Copernican model is equivalent to the Ptolemaic one. The geocentric planetary movements result from vector-sums of the apparent movement of the sun to the heliocentric planetary movements.

In other words: Copernicus could not support with physics his claim to represent the true *forma mundi*, and as a result his system remained an hypothesis in the sense given by the astronomical tradition, and moreover one that was explicitly intended to rehabilitate the principles of the "old," that is Greek astron-

14 See Simplikios (Simplicius of Cilicia), *In Aristotelis physica commentaria*, vols. I-II, ed. H. Diels, Berlin: Reimer 1882/1895 (*Commentaria in Aristotelem Graeca*, vols. IX/X), vol. II, pp. 290 – 291. See also J. Mittelstrass, *Die Rettung der Phaenomene: Ursprung und Geschichte eines antiken Forschungsprinzips*, Berlin: Walter de Gruyter 1962, pp. 140 – 197. And *idem*, "Die Kosmologie der Griechen," in: J. Audretsch and K. Mainzer (eds.), *Vom Anfang der Welt: Wissenschaft, Philosophie, Religion, Mythos*, Munich: C. H. Beck 1989, pp. 40 – 65, 208 – 210, also in: J. Mittelstrass, *Die griechische Denkform: Von der Entstehung der Philosophie aus dem Geiste der Geometrie*, Berlin and Boston: Walter de Gruyter 2014, pp. 43 – 71.
15 *Astronomia nova*, *Gesammelte Werke*, vol. III, p. 25.

omy – the emphasis on circular motions and motions of constant angular velocity. Copernicus believed that these principles had been violated by the previous constructions of kinematic models following the example of Ptolemaic astronomy; his efforts were intended to restore the original principles in astronomy, but on the basis of a heliocentric hypothesis.

So the Copernican "revolution" turns out to be surprisingly conservative: with the intention of changing things in astronomy, Copernicus returns to the Greek beginnings of astronomy, methodologically speaking. This is also why the Copernican propaganda of Giordano Bruno, which stylizes Copernicus not as the founder of a new, but rather as the renovator of an "old philosophy," is not that far off the mark.[16] When examined soberly through the eyes of the history of science, the later and still common presentation of Copernican astronomy in the history of ideas proves to be a misunderstanding. What Copernicus claimed – namely the correspondence of his hypotheses with the cosmological order of the world, or true *forma mundi* – he could not support, for he lacked the physical arguments. And his methodological aims, his strict application of the principles of Greek astronomy, do not lead into a new era but rather into the past.

In light of these considerations one has to ask again: Was Osiander's heresy a betrayal? Probably, only if one keeps in mind that Copernicus rightly considered himself to have been deceived by the preface he had not authorized. With this preface, Osiander foils the author's self-conceptions and aspirations, stabbing him in the back, so to speak. But this attempt – one may be justified in saying – is on the side of scientific truth. It defends truth against exaggerated claims, and resolves a situation before it becomes pure fiction – one, by the way, that the history of ideas will continue working with, staking interpretative claims rather than looking at the scientific facts soberly. This is deception (of the author), then, on behalf of, or in the name of (scientific) truth. A remarkable opposition.

What does this example from the history of science teach us? Certainly not that deception is in some sense normal in science, or capable of being justified on a case-by-case basis; nor that the boundary between scientific truth and scientific deception is fluid. The intention has merely been to show that science is, in theory and practice, more various, more colourful, more complex than even scientific reason itself sometimes imagines. We have reason to insist on an ethos, the ethos of the scientist, which protects science from deception of any kind, and perhaps also from unjustifiable claims to truth.

---

16 *La cena de le ceneri* I, *Le opere italiane di Giordano Bruno*, vols. I-II, ed. P. de Lagarde, Goettingen: Dieterich 1888, vol. I, p. 125.

11 Newton's Concept of Hypothesis and the Origin of Empiricism in Physics

Like every other historical discipline, the historiography of science has its favourite topics. These are related to important discoveries and beginnings that relegate all previous history to the museum of prehistory, and to scientific revolutions that are taken as a starting point for writing the autobiography of modern scientific thought. The distinguishing characteristic of discoveries, beginnings and revolutions for historical purposes is always part of the context of contemporary scientific practice. As opposed to the ideology of pure science such an approach seeks to explain the *historical* character of this practice.

In what follows, we shall deal with a beginning that turned out to be not only a successful *theoretical* beginning (in the sense of a new theoretical concept), but also a beginning of grave *methodological* consequences: Isaac Newton's mechanics, the paradigm of a "new science," as it is called since Galileo Galilei, and his methodology, the paradigm of a new empiricism in science and (parts of) philosophy. The topic is not new, but still not finally clarified.

11.1 Research about hypotheses

Newtonian physics and the history of its diffusion in the eighteenth century is one of those revolutionary turning-points that historians of science are interested in, almost as a matter of course. Apart from the particular scientific gains, Newtonian physics represents, in current terminology, a *paradigmatic change* in the sphere of science, to which also Galileo and Johannes Kepler belong. Through the first comprehensive axiomatic treatment of mechanics, Newtonian physics systematized this scientific practice in a way that served as an important model for the subsequent development of the sciences. The philosopher of science, Hugo Dingler, has summed up Newton's achievement in regard to the establishment of a "theoretical investigation of nature" as follows: Newton saw the point of this research "in the investigation of hypotheses, whereby the noetic forms arrived at mathematically are tested against 'reality,' i. e. by measurements already partially interpreted, and, in the event of agreement, were regarded as 'right.' He is the first true representative of this concept of research, which has dominated the exact sciences to the present day."[1]

1 H. Dingler, *Geschichte der Naturphilosophie*, Berlin: Eidos 1932, pp. 96–97.

https://doi.org/10.1515/9783110596687-011

The distinction between "noetic forms" and their application in a measuring practice already characterized the methodological achievement of Galileo, namely the distinction between a *conceptual* and an *empirical* part of physics. Geometry and kinematics constituted the conceptual part; experiments, the empirical part. With the help of experiments, which are from that time on the major tool of modern science, hypotheses, i.e. propositions of the conceptual part which are linked to existential assertions, are controlled by the introduction of the results of measurements into the formula of the theory.

This is in fact the method which also forms the structural basis of Newton's *Principia*. This structure is developed *synthetically*, i.e. by means of an axiomatic part which adds to the structural development of mechanics already present in Galileo's approach: (1) the concept of force (its dynamic part), (2) the formulation of the law of inertia and (3) the distinction between mass and weight. In regard to this development, Newton's theory of gravity represents the decisive step towards empirical physics. In this theory it is no longer a question of an undefined number of forces and motions, but of one single force, the force of gravity, being the cause of a particular acceleration. The *hypothetical* character of this theory of gravity is most clearly revealed by the fact that its very derivation requires an *empirical* proposition, namely Kepler's Third Law. According to this law the squares of the orbital periods are as the cubes of the radii ($T^2 \sim r^3$) (the course of the planets being approximated by circles). In contrast to the propositions of the axiomatic part, this proposition from the *Principia* is subject to *falsification*. In its theoretical and empirical parts, then, Newtonian physics is research about hypotheses. However, and this is the principal point, this assertion completely contradicts Newton's statements about his own method.

These statements not only collapse the distinction between a theoretical and an empirical part of physics. They also throw into question the very notion of an hypothesis, which links both parts methodologically by means of existential assertions and the conception of an experimental method. "*Hypotheses non fingo*" runs the famous sentence in the *Scholium generale* at the end of the third book of the second edition of the *Principia* (1713). It is already anticipated in Query 20 of the *Opticks* in the Latin edition of the *Opticks* (1706) and in Query 28 in the English edition: "the main Business of natural philosophy," maintains Newton, lies in arguing from the phenomena "without feigning hypotheses."[2] In Query 31 the corresponding sentence runs: "Hypotheses are not to be regarded in experimen-

2 *Opticks or a Treatise of the Reflections, Refractions, Inflections and Colours of Light*, 4th edition, London: William Innys 1730 (ed. I. B. Cohen and D. H. D. Roller, New York: Dover 1952), p. 369.

tal philosophy."[3] In regard to the first and second editions of the *Principia*, this verdict about hypotheses did not have any material consequences. There are only some terminological shifts, the origins of which can be traced back to the year 1706.[4] For example, of the nine propositions characterized as hypotheses at the beginning of the third book in the first edition, including five propositions on planets, these five are characterized in the second edition as "*Phaenomena*" and the first three propositions as "*Regulae philosophandi*." The original third proposition is replaced by a new formulation and later, in the third edition of 1726, a fourth proposition on inductive method is added. The fourth proposition alone, which states that the centre of the entire system is at rest, is still described as a "hypothesis," although Newton placed it elsewhere.

In contrast to the minimal consequences drawn by Newton himself from his verdict about hypotheses, which appeared as a postscript to the original version of the *Principia*, this verdict has had an enormous influence on how Newton has been interpreted and has led to innumerable comments and attempted explanations. The reason is clear: Either Newton's verdict about hypotheses completely contradicts his own method in so far as this is realized in the *Principia*, or one has to go beyond Newton's own statements and infer additional distinctions from the texts. This second approach involves a more differentiated concept of hypothesis. Historical research has chosen unanimously the second possibility, i.e. to show that Newton's own use of the expressions "hypothesis" and "hypothetical" and the method which they articulate are not affected by the verdict about hypotheses in the *Scholium generale*. This corresponds to the historian's customary confidence in the consistency of the texts. In contrast to that, a decision should be made in favour of the first possibility, i.e. that there actually is a conflict between Newton's method and his own methodological or scientific assessment. The reason for this is not some sort of *esprit de contradiction*, but because, hopefully, we are dealing here with a misunderstanding on Newton's part which has not been recognized as such. This is because most interpretations share Newton's own deeper-lying misconception, which will be dealt with later under the heading "empiricism."

The best work in this context is still represented by Alexandre Koyré, I. Bernard Cohen, Alister C. Crombie and Norwood Russell Hanson. Cohen lists 8, Hanson 4 and Crombie 2 systematically defensible meanings of "hypothetical" that differ from Newton's negative verdict about hypotheses, which they treat as a

3 *Opticks*, p. 404.

4 See I. B. Cohen, *Franklin and Newton: An Enquiry into Speculative Newtonian Experimental Science and Franklin's Work in Electricity as an Example Thereof*, Cambridge Mass.: Harvard University Press 1966 (Philadelphia: American Philosophical Society 1956), p. 584.

piece of polemical writing. The difference in the number of positive interpretations among these authors (Koyré only distinguishes between one positive and one polemical case) is quite revealing. Cohen, for example, includes the mathematical case of an unproven (but used) proposition and the axioms on motion formulated by Newton in the first book of the *Principia* under the concept of a (permissible) hypothesis. Hanson, for systematic reasons, first distinguishes four different concepts of hypothesis and subsequently relates them in a not very thorough way to Newton's use of language. Crombie, finally, only distinguishes between the explicit case in which the expression "hypothesis" is preserved and used (as, for example, in the proposition on the centre of the system) and the heuristic use of the term (as, for example, the hypothesis about ether in the *Opticks*). What all of these approaches have in common is to account for the terminological inconsistencies in the Newtonian texts by distinguishing between the verdict about hypotheses as a special polemical case and all other cases in which Newton deals with this issue.

The situation changes, however, as soon as these concepts are confronted with a thoroughly negative statement such as "*hypotheses non fingo.*" According to Crombie and Koyré, "hypotheses" are meant here to refer to mere fictions which cannot be proved. Crombie, for example, sees them as "illegitimate fictions proposed ad hoc and uncontrolled by experimental tests."[5] For Hanson it represents "an expression of some philosophical or metaphysical prejudice."[6] Cohen, finally, in the terms used at the time[7] refers to philosophical fairy tales, "Philosophical Romance" when he speaks of "an hypothesis that produces an arbitrary theory in the absence of experience."[8] In this context it is even possible to bring in a linguistic argument, first noticed by Koyré and later brought into the discussion about Newton's controversial concept of hypothesis by Koyré and Cohen. In the English text of the *Opticks*, "*hypotheses non fingo*" should be read as "I *feign* no hypotheses" and not as has usually been the case ever since Andrew Motte's English translation of the *Principia* in 1729 as "I *frame* no hypotheses." A mistranslation of "*fingo*," which led to the loss of the critical nuances of the sentence, was, so it seems, responsible for the difficulties in interpretation.

---

5 A. C. Crombie, "Newton's Conception of Scientific Method," *Bulletin of the Institute of Physics* 8 (London 1957), p. 360.

6 N. R. Hanson, "Hypotheses fingo," in: R. E. Butts and J. W. Davis (eds.), *The Methodological Heritage of Newton*, Oxford: Blackwell 1970, p. 32.

7 I. B. Cohen, *Franklin and Newton*, pp. 133–134.

8 I. B. Cohen, *Franklin and Newton*, p. 139.

And yet these difficulties are not solved by this explanation. The fact that hypotheses can be *made* (as happens in the *Principia* and the *Opticks*) but that they should not be *feigned* or asserted speculatively does not, in the manner in which it is presented as the actual motive of Newton's verdict about hypotheses, provide any permissible means of distinguishing in a particular case, and in a methodological way, between a permissible and an inpermissible assumption. This would simply mean that the verdict about hypotheses would apply to some hypothetical propositions which one *assumes* to have been arrived at in a conceptually unclear way or which cannot be confirmed by means of empirical method. It would not apply to other propositions where one *assumes* that this is not so. If the positive case is not expressly restricted to cases of direct empirical confirmation that have either been already achieved or can be anticipated on the basis of the experiments carried out, something which Newton himself does not do until well into the 1790s, then the question as to when one can speak of permissible and inpermissible hypotheses remains open.

11.2 Methodological background

The prehistory of Newton's verdict about hypotheses in the second edition of the *Principia* shows that he himself was obviously not aware of this question and that the gradually developing dislike of hypothetical assumptions in his work was rather the product of disputes in which his opponents often made use of the argument relating to hypotheses. In the history of astronomy, this argument acquired a constructive but non-committal character, which seemed to hinder rather than promote the real interests of early modern science: namely in the explanation of real processes by means of mechanical models. In a scholium at the end of the second book of the *Principia*, Newton himself speaks of the Copernican system as the "Copernican hypothesis" (*secundum Hypothesis Copernicaeam*),[9] and he contrasts it with the "*Hypothesis Vorticum*" of René Descartes that is incompatible with the "astronomical phenomena."

The history of this verdict can be traced back to remarks made in discussions, particularly to those found in some contributions to the *Philosophical Transactions* from 1672 onwards. For example, in reply to objections made by Gaston Pardies about the hypothetical (by which is meant here empirically unconfirmed) character of some of his optical propositions, Newton writes: "My design was quite different, for it seems to contain only certain properties of light,

9 *Philosophiae Naturalis Principia Mathematica*, London: Streater and Smith 1687, p. 383.

which, now discovered, I think easy to be proved, and which if I had not considered them as true, I would rather have them rejected as vain and empty speculation, than acknowledged even as an hypothesis."[10] "Vain and empty speculation" is in fact the accusation which Newton levels at the cosmological propositions of Descartes, although without at this stage formulating a general rejection of hypotheses. Newton expresses himself even more clearly in a criticism made in 1672 of Robert Hooke's hypothesis about the spectrum: "whatever be the advantages or disadvantages of this *Hypothesis*, I hope I may be excused from taking it up, since I do not think it needful to explicate my Doctrine by any *Hypothesis* at all."[11]

In a letter sent to Heinrich Oldenburg and forwarded to Pardies, which is printed in volume VII of the *Philosophical Transactions*, Newton attempts, in this connection, an initial methodological definition of the hypothetical procedure: "the best and safest method of philosophizing seems to be, first to inquire diligently into the properties of things, and establishing those properties by experiments and then to proceed more slowly to hypotheses for the explanation of them. For hypotheses should be subservient only in explaining the properties of things, but not assumed in determining them."[12]

All this would probably have remained at the level of casual comments directed to particular cases against unjustifiable speculative assumptions, had Newton not thought it necessary to carry his dispute with Cartesian physics beyond controversies of fact to the *methodological* level. This dispute can be traced through all of Newton's published and unpublished works. It became increasingly methodological and theoretical once the superiority of Newtonian physics over Cartesian physics was clearly recognized not only in England, but also on the continent through the French translation of the *Opticks* in 1720 and the work of Peter van Musschenbroek, Willem s'Gravesande and Pierre Maupertuis. As early as 1697 Jacques Rohault's influential *Traité de Physique* (1671), a standard textbook of Cartesian physics, was published in England in a Latin translation (and then in 1723 in an English translation) together with a critical commentary which included references to Newtonian physics. This is without doubt a scientific curiosity, but it does show that even at this early date Cartesian physics was read through the eyes of Newtonian mechanics, at least in England. As far as the often mentioned Cartesian theory of vortices is concerned, i.e. the explana-

---

10 *Isaac Newton's Papers and Letters on Natural Philosophy and Related Documents*, ed. I. B. Cohen and R. E. Schofield, Cambridge: Harvard University Press 1958, p. 92, cf. p. 109.

11 Ed. I. B. Cohen and R. E. Schofield, p. 123.

12 Ed. I. B. Cohen and R. E. Schofield, p. 106; cf. *The Correspondence of Isaac Newton* I, ed. H. W. Turnbull and J. F. Scott, Cambridge: Cambridge University Press 1959, p. 164.

tion of the movement of the planets as produced mechanically by an ether vortex, this was immediately recognized as being inferior to the Newtonian theory of gravity and its dynamic derivation from Kepler's laws. It is not this inferiority, however, but the verbal disqualification as a "hypothesis" which Newton uses to disqualify the vortex theory. In doing so he causes first terminological and then methodological difficulties for himself.

From the draft of a fifth "*Regula philosophandi*" discovered among the manuscripts by Koyré, it is clear that Newton deliberately chose not to restrict the controversy with Descartes to the sphere of physical theorems, but to extend it to the methodological and epistemological spheres. In this rule Newton uses the terminology of John Locke to criticize Descartes' rationalism, in so far as it takes *cogito ergo sum*, i. e. the construction of a mental Archimedean point, and the conception of so-called innate ideas (*ideae innatae*) as its starting point. Such a criticism, which would have made Newton, at least for a time, into an ally of Gottfried Wilhelm Leibniz (which was certainly not intended) does not fit into the argumentative context of the *Principia* (and was probably omitted from the "*Regulae*" for this reason). But it does reveal that the "*hypotheses non fingo*," whatever difficulties it causes within a physical and methodological context, is part of a far-reaching strategy.

This strategy can be seen as an attempted defence against the epistemological tutelage of Locke's philosophy which became increasingly evident after the publication of the *Essay concerning Human Understanding* (1690). Newton himself now undertakes the theoretical argument against Cartesian metaphysics, which he rightly regards as having been inadequately addressed in Locke's epistemological writings. It is for this reason that he starts with the concept of hypothesis, which was a central feature of the theoretical and methodological discussions of the time. But this concept was not really controversial and it was thoroughly understandable in its use. The accepted terminology is summed up in 1704 by John Harris in his influential *Lexicon Technicum* (the same lexicon whose fifth edition in 1736 registers contemporary difficulties with regard to the understanding of "Newtonian Philosophy"). Five meanings are given. Harris writes: "Hypothesis, the same with *Suppositions*. When for the Solution of any *Phaenomena* in Natural Philosophy, Astronomy, etc. some Principles are supposed as granted, that from thence an Intelligible and Plausible account of the Causes, the Effects of the proposed *Phaenomena* may be given, the laying down or supposing such Principles to be granted, is called an *Hypothesis* (...). Wherefore an *Hypothesis* is a *Supposition of that which is not, for that which may be*; and it matters not whether, what is supposed be true or not, but it must be possible, and should always be probable."

There can be no doubt that Newton's own method up to this point corresponds to this definition (Harris follows this up with a reference to the synonymity in astronomy of "system" and "hypothesis"). But it is clear as well that Newton's verdict about hypotheses, which is also understandable in its intention, brings with it the risk that he can no longer draw the line between permissible and inpermissible hypotheses with the necessary clarity. This is definitely true of the reception of Newton in the following years. In a book published in 1735, *The Philosophical Grammar: Being a View of the Present State of Experimented Physiology, or Natural Philosophy,* Benjamin Martin, who, among other things, wrote *A Plain and Familiar Introduction to the Newtonian Philosophy* (1751), gives the following illuminating answer to the question: "Are any Kinds of Hypotheses to be admitted in reasoning about natural Subjects?": "The Philosophers of the present Age hold them in vile Esteem, and will hardly admit the Name in their Writings; they think that which depends on these Hypotheses and Conjecture, unworthy the Name of Philosophy."[13] Newton's verdict is making history, then. A history that Newton certainly did not intend in regard to its naive generalization, but which can scarcely, on the other hand, be presented as a simple misinterpretation of his statements. The verdict is too obviously present in Newton's work for that. This remains the case even if one takes into account the fact that the verdict not only is *offensive* in character but also serves the purpose of a *defensive* statement.

As far as the defensive character of the verdict is concerned, Leibniz, in a letter written on February 10th, 1711 to Nicolaus Hartsoeker, accuses Newton, without mentioning his name, of reintroducing occult qualities into physics with his concept of gravity. Roger Cotes, the editor of the second edition of the *Principia*, informed Newton on March 18th of this accusation, which had in the meantime been published, and recommended that it be taken into account in the *Principia* as follows: "I do not propose to mention Mr. Leibniz's name, twere better to neglect him, but the Objections I think may very well be answered & even retorted upon the maintainers of Vortices."[14] Precisely this, the readdressing of Leibniz' accusation, occurs in the *Scholium generale.* Newton here defends his concept of gravity against methodologically oriented attacks. In the *Principia*, the explanation of gravity constitutes the immediate context for the passage on hypotheses; in the *Opticks* Newton wants to distinguish gravitation – as well as

---

13 B. Martin, *The Philosophical Grammar: Being a View of the Present State of Experimented Physiology, or Natural Philosophy,* London: J. Noon 1735, p. 22 (7th edition, London: Rivington etc. 1769, p. 19).

14 *Correspondence of Sir Isaak Newton and Professor Cotes, Including Letters of Other Eminent Men,* ed. J. Edleston, London: John W. Parker and Cambridge: John Deighton 1850, p. 153.

the other two active principles (fermentation and cohesion) – from methodologically suspect occult qualities inspired by Aristotelian physics.[15] These qualities had already been rejected by Robert Boyle. As a result Newton had to be careful not to get caught in the line of fire of his own supporters. His defensive strategy was to privilege gravitation as an empirically provable phenomenon, its explanation being desirable but not necessary.

11.3 Observation and experiment

However clearly we present the reasons which led Newton to his verdict about hypotheses, this is not the end of the matter either for Newton or for the interpretation of Newton. With his verdict, Newton intended to strike a blow at speculative science (and to protect his own work from the accusation that it was speculative), but at the same time it also affected his own understanding of methodology in a way which was to have serious consequences. If we take the verdict in the form in which it first appears in Query 20 of the Latin *Opticks* of 1706 and later as the assurance "without feigning hypotheses" in the revised English version of the *Opticks* of 1717/18 (now as Query 28), it becomes a central part of a methodology that begins to move further and further away from the actual structure of the *Principia*. The tendency of these theoretical reflections, which are in such strange contrast to this structure, cannot be overlooked. It can be characterized as a first attempt to develop an *empiricist foundation* for physics.

Nothing in the early versions of the *Principia* suggests this later development. In a scholium in the first book of the first edition of 1687 the physical method is described as part of a mathematical method. In mathematics (*in mathesi*), writes Newton, one has to investigate the quantities of forces (*virium quantitates*) and those relationships between them (*rationes*) "which follow from certain assumed conditions. If one then descends to physics (*deinde ubi in Physicam descenditur*), one must compare these relationships with the phenomena in order to find out which conditions of the forces correspond to the particular kinds of attractive bodies."[16] If one translates "*in mathesi*" with "in the theoretical part of physics" and "*in physicam*" with "in the empirical part of physics", this corresponds precisely to the insight which forms the basis of Galilean kinematics and which was referred to above under the concept of "investigation of hypoth-

15 *Opticks*, p. 401.
16 *Principia*, p. 192.

eses." This is that the structure of proof in physics requires a synthetic element characterized by conceptual parts, which then permits the determination of the actual empirical part by means of empirical existential assertions and their experimental control. What Newton expresses by means of metaphor ("descend") refers in methodological language to the (correct) sequence of steps which is constitutive of physics.

The decisive point here is that Newton's own words describe precisely the method pursued in the *Principia*. This work contains an axiomatic part, with the help of which particular theorems are derived. In contrast to this foundation of theoretical dynamics built upon the three fundamental quantities of length, time, and mass, there is then an empirical part in which existential assertions such as the assumption of a general attractive force (which is used in order to derive Kepler's elliptical orbits of the planets) play the methodological role of transition. Methodological considerations about the kind of explanation given are on the whole of secondary importance, but when they do occur, they confirm the procedure chosen in practice.

Reflection on methods is completely different in so far as it occurs in the *Opticks* and in the second edition of the *Principia* of 1713. The catchword is now "experiments and observations;"[17] propositions in physics of any kind whatsoever (although, in the background, doubtless, the law of gravity, itself an empirical proposition, plays an exemplary role) are regarded as "deduced from Phaenomena and made general by Induction."[18] This is now asserted even for the axioms at the beginning of the first book of the *Principia* with recourse to the vexatious question of hypotheses: "in Geometry the word Hypothesis is not taken in so large a sense as to include the Axiomes & Postulates, so in Experimental Philosophy it is not to be taken in so large a sense as to include the first Principles or Axiomes wch I call the laws of motion. These Principles are deduced from Phaenomena & made general by Induction: which is the highest evidence that a Proposition can have in this Philosophy."[19] Propositions which Newton had formerly characterized as "hypotheses" now appear as empirical generalizations and thus as propositions of the kind that Newton had derived empirically in the third book, although, here too, he makes use of the entire deductive apparatus provided by the theoretical dynamics of the first two books. For methodological reasons, the physics of the *Principia* is reduced, at this late point in time, to

---

17 *Opticks*, p. 404.
18 Ed. J. Edleston, p. 155.
19 Ed. J. Edleston, pp. 154–155.

its empirical part. At the same time, without Newton's noticing, his theory comes into conflict with is own methodological presentation.

11.4 Analytical and synthetical method

If one looks for a coherent account of the reflection on method, as begun later by Newton without changing anything in the inner structure of his physics, a passage from Query 31 of the *Opticks* suggests itself. It reads: "As in Mathematicks, so in Natural Philosophy, the Investigation of difficult Things by the Method of Analysis, ought ever to precede the Method of Composition. This Analysis consists in making Experiments and Observations, and in drawing general Conclusions from them by Induction, and admitting of no Objections against the Conclusions, but such as are taken from Experiments, or other certain Truths. For Hypotheses are not to be regarded in experimental Philosophy. (...) And if no Exception occur from Phaenomena, the Conclusion may be pronounced generally. But if at any time afterwards any Exception shall occur from Experiments, it may then begin to be pronounced with such Exceptions as occur. By this way of Analysis we may proceed from Compounds to Ingredients, and from Motions to the Forces producing them; and in general, from Effects to their Causes, and from particular Causes to more general ones, till the Argument end in the most general. This is the Method of Analysis: And the Synthesis consists in assuming the Causes discover'd, and establish'd as Principles, and by them explaining the Phaenomena proceeding from them, and proving the Explanations."[20]

These propositions describe the relationship between two methodical procedures, the *analytical* and *synthetical* methods, which were in the future to determine the methodological approach in physics and, subsequently, in all of the sciences. The fundamental role of the analytical method in this methodological concept, i.e. the priority of inductive over synthetic procedures, is formulated in a manner which was current in the contemporary reflections on method, but had a completely different meaning from than. Starting from the logical method of proof in the Aristotelian *Second Analytics*, Paduan Aristotelianism had already introduced the distinction between a *metodo risolutivo* and a *metodo compositivo* in order to find a basis for empirical science. This distinction was later adopted by Galileo in his statements on methodology. In its Galilean form *metodo risolutivo* describes the way in which one arrives at explanatory propositions, whereas the *metodo compositivo* is reserved for the formulation of hypotheses, which are

20 *Opticks*, pp. 404–405.

then drawn upon by means of a renewed use of the *metodo risolutivo*. The decisive point, however, is that this methodological concept still represents a logical method of proof in which the first step, the *metodo risolutivo*, is not empirical, but remains a conceptual step (taken within an axiomatic approach).

The situation is completely different in Newton's case. The analytical method does take the place of the *metodo risolutivo*, but now even this step, the formulation of causes for observed effects, is subject to empirical control. What Newton describes as "analysis" is not a conceptual or an *a priori* provision of distinctions and principles, but a prescription to begin one's investigation with the phenomena themselves and by means of induction, framed in general (empirical) propositions. The fact that the step from motions to the (motive) forces, i. e. the step from the effects to the causes, is provided for, makes it clear that the whole process represents an attempt to find a basis for a methodology of cause and effect in which the older synthetic part merely signifies the conclusion about effects from *observed* causes.

This makes clear, once again, what Newton means when he says that the axioms of his mechanics are "deduced" from phenomena. According to Newton's interpretation, and in contrast to former interpretations in the tradition of Galilean physics, these axioms are empirical generalizations and thus subject to empirical control. This consequence, actually, does not appear in this form in Newton's work, and there is also no indication whatsoever as to how such propositions, for example, the law of inertia, are to be derived "from experience." But all of this seems unproblematical for Newton. The same applies to the methodologically orientated Newtonian tradition, which took the verdict about hypotheses as literally as it could only really be taken by someone who had never read the *Principia*. Thomas Reid, the founder of the Scottish School, still describes Newton's methodological achievement with the following two directions: "first, by just induction from experiment and observation, to discover the laws of nature; and then, to apply those laws to the solution of the phaenomena of nature."[21] And John Harris, embarrassed by having to place various meanings of Newtonian Philosophy alongside one another, notes in the fifth edition of his *Lexicon* of 1736 the following interpretation current at that time: "Others by Newtonian Philosophy, mean the Method or Order which Sir Isaac Newton observes in Philosophising, viz. the Reasoning, and drawing of Conclusions directly from Phaenomena, exclusive of all previous Hypotheses; beginning from

---

21 Th. Reid, *Essays on the Active Powers of Man* (Essay I, Chap. VI), *The Works of Thomas Reid*, vols. I-II, ed. W. Hamilton, Edinburgh: MacLachlan, Stewart, and Co. and London: Longman, Brown, Green and Longmans 1846/1863, vol. II, p. 527.

simple Principles; deducing the first Powers and Laws of Nature from a few select Phaenomena, and then applying those Laws, etc. to account for other things."

Any number of examples illustrating this view could be given, for example in the works of Colin Maclaurin, William Emerson, Henry Pemberton, Willem s'Gravesande, John Keill, John Martin, Voltaire and Pierre Louis Moreau de Maupertuis. It is not surprising that Newton now appears as a model pupil of Francis Bacon and thus as a representative of a philosophy of science which at the time of its conception had already been left behind by the methodological turning point in science established with Kepler and Galileo. Maclaurin writes: "Sir Francis Bacon Lord Verulam who was contemporary with Galileo and Kepler is justly held amongst the restorers of true learning, but more especially the founder of *experimental philosophy.*"[22] Newton accepted this genealogical reconstruction of his own methodological position, as it was presented, for example, by Pemberton, the editor of the third edition of the *Principia*, who made it the basis of his account of Newton's achievement in his book *A View of Sir Isaac Newton's Philosophy* (1728). Newton preferred to be methodically close to Bacon rather than to Descartes.

Newton's momentous methodological judgment of his own achievement led to an historical accident in the development of modern physics, which, in the final analysis, can be traced to the fact that Newton unthinkingly allowed the framework of the debate to be prescribed by his opponents. Newton forcefully opposes a speculative tendency which he sees developing again, but because he identifies this tendency from the very start with the position of rationalism, of Cartesian philosophy, he overlooks the fact that this position can be understood differently, as it sometimes is, for example, in Leibniz. In his dispute with Leibniz, which is itself impaired by numerous misunderstandings, Leibniz' position appears, so to speak, as "exhausted" – so that the synthetic structure of physics in Galileo, Huygens and Newton himself is suspected by their critics of sharing Cartesian positions in methodology in spite of its non-Cartesian content. And for this reason it is reinterpreted. In the third edition of the *Principia* (1726) the newly added "Fourth Regula Philosophandi" runs: "In experimental philosophy we are to look upon propositions collected by general induction from phenomena as accurately or very nearly true, notwithstanding any contrary hypotheses that may be imagined, till such time as other phenomena occur, by which they may either be made more accurate, or liable to exceptions. This

22 C. Maclaurin, *An Account of Sir Isaac Newton's Philosophical Discoveries, in Four Books*, ed. P. Murdoch, London: A. Millar etc. 1748, p. 56 (2nd edition London: A. Millar 1750, p. 59).

rule we must follow, that the argument of induction may not be evaded by hypotheses."

In this rule, which sums up once again the methodological statements of the *Opticks*, physical theory is finally put on a basis which, although fragmentary and not at all suited to this theory, leads directly to the empiricism of modern science. The assumption, central to classical empiricism, that physical data are empirically "given in pure form" and cannot be falsified by any system of conceptual distinctions – this assumption is here taken up for the first time in opposition to an elaborated physics. It corresponds to the epistemological assertion, first made by Locke, of a conceptfree basis of all knowledge in experience. Both Newton and the philosophical representatives of classical empiricism overlook the fact that, although such a formulation is contrary to the rationalist position of Cartesian philosophy, it does paradoxically adopt a rationalist assumption, namely the assertion that the (physical) world "itself" is well-ordered. In orthodox empiricism the rationalist assertion of a *mental* anticipation of the "structure of reality" is simply replaced by the construction of an *empirical* sensorium: observation replaces meditation, the statements about the (physical) world with regard to its recognizable structure remain the same, even when the physical propositions change.

11.5 Empirism misunderstood

This makes it clear, once again, that the methodology subsequently added to Newtonian physics was superimposed upon an achievement which it did not really fit, and in a way that can only be explained historically, not systematically. The attempt to relate this methodology to Newtonian physics by means of a more extensive systematization, fails. It fails because it cannot explain the structure of theoretical dynamics in the *Principia*, i.e. the *Euclidian* character of a *rational* mechanics cannot be understood according to it. Hence Paul Feyerabend's suggestion that Newton's term "deduced from the phenomena" is to be understood as a derivation already based upon theoretical distinctions. According to Feyerabend, these phenomena are not meant to refer to "everyday facts pure and simple," but to "an intimate *synthesis of laws*"[23] or to "an idealized and generalized

23 P. K. Feyerabend, "Problems of Empiricism," in: R. G. Colodny (ed.), *Beyond the Edge of Certainty: Essays in Contemporary Science and Philosophy*, Englewood Cliffs: Prentice Hall 1965, pp. 159–160.

description of the result that uses the terms of the theory under review."[24] In this suggestion, in other words, everything is already included in the concept of physical phenomena that is later to form its theoretical relationship with a synthetic structure of physics. But this means that Newton would now be presented with the same difficulty that he believed he had avoided by returning to an inductive method as against the conceptual assumptions of the Cartesian position, namely the task of demonstrating what distinctions have already entered into the establishment of physical processes in order that these can be grasped as a first phase of physical explanation.

That such an approach is possible is first shown by Immanuel Kant. Newton himself, however, was probably of the opinion that every step in this direction would mean a step back to the speculative tradition of "natural philosophy" of the old kind and to Cartesian metaphysics. At any rate, there is no evidence for a positive judgment on the possibility of such a step. And Feyerabend's suggestion is presumably based, in essence, upon the fact that in the second edition of the *Principia* Newton renamed some of the propositions formerly called "hypotheses" as "phenomena."

In spite of this suggestion, Feyerabend also sees in Newtonian physics the beginning of the tradition of classical empiricism, which, as the "official doctrine" of classical physics, made the assumption of a "stable experience" the basis of methodological distinctions. This does not mean that empiricism succeeds in developing the modern structure of physics from the fragmentary Newtonian methodology. The peculiar discrepancy between physical practice and an empiricist methodology remains even in the future broadly characteristic of the development of classical and of modern physics. In the analytical systems of Leonhard Euler, Joseph Lagrange, Jean Baptiste Joseph Fourier and James Maxwell, for example, Newton's axioms of motion are replaced by systems of equations, i. e. axiomatic and deductive elements are replaced by simple calculations. These are then described as an "explanation" of physical processes when a series of measurements which are already available can be derived from them in accordance with an adequate mathematical description.

The assertion that the methodology of empiricism is then satisfied, because a beginning which is free of theory has been made with a series of measurements, i. e. with empirical data, this assertion is only apparently justified. In the analytical systems the question as to how the "first" propositions of a Newtonian theory are to be arrived at is indeed disposed of. However, the theoretical

---

24 P. K. Feyerabend, "Classical Empiricism," in: R. E. Butts and J. W. Davis (eds.), *The Methodological Heritage of Newton*, Oxford: Blackwell 1970, pp. 106–107.

approach which is the basis of the apparatus contains *synthetic* elements. These elements include the determination of the basic concepts of a theory, in the case of dynamics, for example, the determination of the concepts of impulse and kinetic energy, without which the transition from mathematical statements to physical statements, i.e. the physical interpretation of the algorithms of analytical theories, would not be possible. In other words, the observance of an empiricist methodology is bought at the expence of a synthetic deficit that is only recognized much later, beginning with Heinrich Hertz and Ernst Mach.

The controversial role which empiricist concepts in science have played and still play can thus be traced back to the beginnings of a methodology that Newton developed *for defensive purposes* as an explanation of his own achievements. It is the combination of Newtonian theory and Newtonian methodology which does not really fit the theory that forms the basis of modern empiricism in science. From an historical point of view, it is only as a result of this development that the Baconian distinctions, which lay completely outside the mainstream of thought, gained significance and that epistemological empiricism in the tradition of Locke gained the upper hand over a rationalism which could not escape from its Cartesian past. That rationalism, however, was much closer to Galilean and Newtonian physics in regard to its synthetic character than empiricist propaganda would have us believe. A scientific practice which had begun successfully, the practice of Galilean and Newtonian physics, is thus retracted theoretically and has to justify itself against a methodology to which it is completely opposed. A truly paradoxical situation. As this analysis of Newton's thought shows, this kind of empiricism is in part a product of a philosophical misunderstanding.

12 Philosophical Foundations of Science in the 20th Century

The 20th century was an important century in the history of the sciences. It deserves to be called a *scientific century*. It generated entirely novel insights in foundational issues and established a previously unknown intimate connection between science and technology. Whereas physicists at the end of the 19th century had thought of themselves as having reached the end of basic research and had believed the principles of physics to have been discovered in their entirety, in the first third of the 20th century we witness revolutionary changes, comparable to the scientific revolution of the 17th century.

With the development of the Special and the *general theory of relativity* as well as quantum theory, the central theoretical frameworks of modern, non-classical physics were introduced. Theoretical investigations into the statistical interpretation of thermodynamics and infrared radiation lead to the development of quantum mechanics, which in turn prompted modifications of the atomic model and allowed an explanation of the photoelectric effect. The development of the *special theory of relativity* as a theory of the spatio-temporal relationships between inertial systems moving relative to each other, which yields an explanation of the properties of transformations of the Maxwell-Hertz equations, and of the General Theory of Relativity as theory of the classical (non-quantised) gravitational field, leads to entirely new conceptions of space, time and gravity. Essential steps in the development of quantum mechanics are the development of quantum statistics and of the uncertainty principle which sets limits on the measurement of atomic processes. In contrast to classical physics, natural laws preclude determinate measurements of the system's state. At the same time, essential clarifications and specifications are made to fundamental concepts of epistemology (or natural philosophy) such as the concepts of space and time in the *theory of relativity*, of causality and locality in *quantum theory*, of matter and field in the physics of elementary particles.

Besides physics, the discipline of biology, especially molecular biology and biophysics, which, together with biochemistry, conceives of itself as a molecular research programme, as well as evolutionary theory, become a leading science. Within biology, due to the discovery of the chemical structure of the DNA and the deciphering of the genetic code, the 20th century has been called the century of

https://doi.org/10.1515/9783110596687-012

the gene.[1] Developments in other parts of the natural sciences, such as astrophysics, chemistry, in the earth and environmental sciences as well as in the neurosciences are of comparable significance. In addition, there is an ever closer connection between science and technology. Scientific research has reached a point where idealizations may be overcome and the controlled laboratory may be left behind. Rather, science is now in the position to do justice to the complexity of the real world.

These developments are accompanied by *epistemological* reflections. On the one hand, these are directly connected to the scientific developments and, as in the case of the concepts of space and time, are part of scientific theory construction; on the other hand, general philosophy of science experiences an increase in importance and influence within that part of philosophy which is close to science. Science does not just yield important discoveries, it also becomes reflexive – in the sense of making its own procedures, theoretical, methodic and empirical, the subject of critical scrutiny. This is especially true concerning the foundations of science.

In what follows, a few brief remarks on the topic of philosophical foundations will be made, addressing three different *epistemological* approaches: one that is scientific in the narrow sense, emerging out of scientific theorizing itself, one that is both scientific and philosophical (mediating, in a sense, between science and philosophy), and one that is of a general philosophical nature (general in the sense of *general philosophy of science*). They are all representative of the connection between science and epistemology, and they all illustrate the high standard of scientific thought in the 20th century. To conclude, a few remarks on developments relating to new forms of organizing research and a revised concept of research follow.

1. An approach that is scientific in the narrow sense is connected to epistemological problems which are primarily of scientific importance. Questions raised by quantum mechanics belong to this area. In the so-called Copenhagen interpretation, a correspondence principle bridges the gulf between classic and quantum-theoretic explanations of the structure of matter. At the same time, the differences between quantum mechanics and classical physics lead to different epistemological interpretations, for instance an *instrumentalist* reading, according to which quantum mechanics is not about the physical reality as such, but about a world as perceived by the epistemological view of the physicist, or a *realist* interpretation, for instance that advocated by Albert Einstein, according to

1 E. F. Keller, *The Century of the Gene*, Cambridge Mass. and London: Harvard University Press 2000.

which the physical objects exist independently of each other and the context of measurement.

An instrumentalist approach also implies the view that there are principled epistemological limits to knowledge or human cognition, whereas a realist approach implies the (problematic) view of the incompleteness of quantum mechanics, which might be overcome by assuming hidden parameters. Other examples might be the issue of the conventional nature of simultaneity within *special relativity* and the debate in the foundations of mathematics, in which formalist, platonist and constructivist conceptions were competing as the bases of mathematics.

2. Connected to epistemological problems of this kind, resulting directly from scientific research, are ones of scientific as well as of philosophical significance. Among these are, for instance, the topics of determinism, emergence, and (again) realism. Everything we know about the world, in science and philosophy, seems to depend on the question whether we live in a *deterministic world.* A well-known example for this is chance in quantum mechanics.[2] Quantum mechanics imposes serious limitations on the predictability of events. The central principle of the theory is *Schroedinger's equation*, which serves to determine the "state function" or "wave function" of a quantum system. The state function is generally taken to provide a complete description of quantum systems; no properties can be attributed to such a system beyond the ones expressed in terms of the state function. *Schroedinger's equation* determines the time development of the state function unambiguously. In this sense, quantum mechanics is a *deterministic* theory.

However, apparently irreducible chance elements enter when it comes to predicting the values of observable quantities. The measurement process in quantum mechanics is described as the coupling of the quantum system to a particular measuring apparatus. *Schroedinger's equation* yields, then, a range of possible measuring values of the quantity in question, each of these values being labelled with a probability estimate. That is, *Schroedinger's equation* only provides a probability distribution and does not anticipate particular observable events. Heisenberg's so-called indeterminacy relations are a consequence of *Schroedinger's equation*, although historically they were formulated independently of this equation and prior to its enunciation. The Heisenberg relations place severe limitations on the simultaneous measurement of what are called "incompatible" or "incommensurable" quantities such as position

2 On this and the following point on "emergence," compare the more extensive treatment in chapter 2.

or momentum or spin values in different directions. The more precisely one of the quantities is evaluated, the more room is left for the other one. Like the constraints mentioned before, the limitations set by the Heisenberg relations have nothing to do with practical impediments to increasing measurement accuracy that might be overcome by improved techniques. Rather, the relations express limitations set by the laws of nature themselves. This element of genuine, irreducible chance troubled Einstein very much. It challenges the thesis of a deterministic world.

Concerning the concept of *emergence*, what is at issue is the relationship of properties of wholes to properties of its component parts, equally relevant in science and philosophy. Originally, it made reference to the conceptual contrast, in a biological context, between "mechanicism" (as a particular variant of materialism) and "vitalism." Systematically, it says that it is insufficient to use characteristics of elements and their interrelations to describe characteristics of ensembles or make predictions about them[3] (the whole is more than its parts[4]). According to the *emergence thesis*, the world is a levelled structure of hierarchically organized systems, where the characteristics of higher-level systems are by and large fixed by the characteristics of their respective subsystems, yet at the same time essentially different. Different characteristics and processes occur in the respective levels. Furthermore, weak and strong emergence theses can be distinguished.

The core element of the strong emergence thesis is the non-derivability or non-explainability hypothesis of the system characteristics shaped from the characteristics of the system components. An emergent characteristic is non-derivable; its occurrence is in this sense unexpected and unpredictable. Weak emergence is limited to the difference of the characteristics of systems and system components and is compatible with the theoretical explainability of the system characteristics. Weak emergence is essentially a phenomenon of complexity. Of scientific interest is particularly the *temporal* aspect of the emergence thesis, i.e. for ensemble characteristics that occur in developments. Limits of reducibility (of the whole to its parts) figure here as limits of explanation and predictability which is an important criterion of a justified theory and thus its achievement. This temporal novelty is described by the concept of *creative advance of nature.*

---

3 For the followings see M. Carrier, "emergent / Emergenz," in: J. Mittelstrass (ed.), *Enzyklopaedie Philosophie und Wissenschaftstheorie*, 2nd edition, vol. II, Stuttgart and Weimar: J. B. Metzler 2005, pp. 313–314.

4 See K. Lorenz, "Teil und Ganzes," in: J. Mittelstrass (ed.), *Enzyklopaedie Philosophie und Wissenschaftstheorie*, vol. IV, Stuttgart and Weimar: J. B. Metzler 1996, pp. 225–228.

All these epistemological reflections, in science as well as in philosophy, are related to the already-mentioned *realism* debate. In philosophy, one distinguishes between two kinds of realism. *Ontological realism* is the position that the world of objects exists independently of human perception, knowledge and thought; *epistemological realism* – in contrast to idealism, which thinks of the world as being a construction of the self or a representation of the world, respectively – is the position that in the process of discovery, the objects of discovery play an independent role. So epistemological realism assumes essential elements of ontological realism, put simply, the existence of an "external world." To the extent that in (philosophical or scientific) theories a realist stand is taken, these are called *empiricist* when they make reference to the relation of the object of discovery and the subject of discovery, or *platonist* when they make reference to the status of general concepts, so-called *universals*. Accordingly, a distinction may be made between empiricist and platonist positions on scientific theory formation.

The status of a theory furthermore depends, also from the epistemological point of view, on the interpretation chosen, also concerning determinism and realism. An example would be the interpretation of the electromagnetic field as a state of a mechanical ether in the mechanistic tradition of the 19th century. Departing from this interpretation, Einstein conceived of this field as an independent magnitude. Both are different (possible) interpretations of the same Maxwellian theory of electrodynamics. Furthermore, it is disputable, whether a relational theory of space, according to which space represents merely a relation among objects and does not itself exist beside the objects or outside them, is really adequate to the *general theory of relativity* – as Einstein himself believed. Depending on how one translates classical relationalism into the concepts of relativity theory, one receives different answers to the question. At the moment at least, it is impossible definitely to privilege one particular one of these translations. In other words: One and the same theoretical approach can be differently interpreted; interpretations in these scientific cases, too, are not unequivocal. On the contrary, they display characteristic uncertainties that cannot be completely removed even by a rational reconstruction of the basic principles underlying a theory. The interpretation of quantum theory is not essentially different in this regard from an interpretation (say) of Immanuel Kant's theory of space and time.

In all of these cases we are dealing with questions and areas of research whose results are not clearly attributed to physics or philosophy. This is well illustrated by physicist-philosophers such as Einstein, who first endorsed an operationalist, later a realist epistemology, or Werner Heisenberg, who pursued the project of finding a *theory of everything*, believing in homogeneous mathematical symmetry, or Stephen Hawking, who writes on quantum cosmology from a general epistemological perspective, endorsing a falsificationist position in the sense of Karl R. Popper.

3. A properly philosophical status may be attributed to epistemological reflections which in the 20th century gained significance as a discipline entitled *philosophy of science*. These in general deal with problems of structure and development of science, starting from a distinction between *research form* and *theory form* of science. In its research form science is trying to discover what is the case, in its theory form it represents what it has discovered. Science in the research form is an expression of object rationality (including questions regarding the constitution of objects), science in the theory form is an expression of rationality in justification. Epistemology in the domain of science essentially refers to the theory aspect, namely to questions regarding the *structure*, *dynamics* and *explication* of theories. Under the heading "theory structure" it analyses the structures of the language of science and of scientific explanations and the formation of theories. Under the heading "theory dynamics" it deals with the developmental structures of scientific theories and with questions concerning the criteria of comparative theory assessment. The heading "theory explication" applies to questions such as "Is there a physical basis for the direction of time?" or "Does the wave function of quantum mechanics refer to individual particles or an ensemble of particles?" (the Copenhagen versus the statistical interpretation). As examples for such forms of thinking about science the influential approaches of Logical Empiricism (Rudolf Carnap being the main representative) and that of Popper may be mentioned.

Logical Empiricism, which epistemologically may be characterized by its appeal to the conventionalism of Henri Poincaré and its criticism of the thesis of the *synthetic a priori* of Kant, conceives of theory development as a continual progress of discovery in which earlier theories are reduced into later ones. Epistemologically speaking, it endorses a two-level view of the conceptual structure of scientific theories, according to which in the structure of science all true propositions are either logically or analytically true propositions, or alternatively empirically or synthetically true propositions.

On this basis, it at the same time pursues the project of the *unity of science:*[5] all states of affairs can be expressed in a physicalist language and by introducing *theoretical concepts*, i.e. concepts which refer to entities not directly observable and which cannot be defined in terms of observational concepts. They are introduced by the postulates of a theory and their function and role is explicated ac-

5 See M. Carrier and J. Mittelstrass, "The Unity of Science," *International Studies in the Philosophy of Science* 4 (1990), pp. 17–31.

cordingly by the appropriate theoretical context. While theoretical concepts are generally coordinated with observational indicators by correspondence rules, nonetheless, these concepts cannot be translated into such empirical indicators. The reason for their introduction is that they help to order and unify experimental laws successfully. Concepts such as electromagnetic field or the quantum-mechanical wave function, to which empirical characteristics can be assigned only indirectly, partially, and in a manner mediated by theory, are considered legitimate, because with their help the explanatory power of the theories can be increased. Theoretical concepts are thus legitimate explanatory constructs. The conceptional structure of scientific theories according to this position is shaped accordingly.

Popper's approach was very different. Opposing the idea of how the reducibility of theories into each other leads to scientific progress in Logical Empiricism, he defends the incompatibility of successive theories. In his methodology of empirical science or *logic of scientific discovery*, entitled "falsificationist," the term "corroboration" takes the place of the concept of justification, in particular, empirical justification, as Popper – again, in opposition to Logical Empiricism – appeals to the asymmetry of verification and falsification: general propositions, mostly natural laws, may only be refuted (falsified), but not verified, relative to an empirical basis. Basic propositions, which according to this conception figure as premises of an empirical falsification, are interpreted as corroborating a falsifiable hypothesis. The degree of corroboration of a theory in turn depends on its degree of testability, expressed by the concept of falsifiability. The principle of a critical examination characterizing a logic of scientific discovery accordingly requires a pluralism of theories so as to be able to select a "successful" one, which later (against Popper) was extended by a pluralism of methods by Paul Feyerabend. Progress among theories is due to the ongoing process of critical revision of existing theories from the perspective of truth or at least verisimilitude.

In his later works, Popper tried to describe the formation of theories as an evolutionary process, as the expansion of knowledge in problem-solving contexts, the components of which are creative guesswork and the rational elimination of error. This process is supposed to be based on a "third world of objective contents of thought," existing alongside the "first world" of physical objects and the "second world" of mental states. Opposing this we find *historicist* approaches (Thomas S. Kuhn), *reconstructivist* approaches (Imre Lakatos), *structuralist* approaches (Joseph Sneed, Wolfgang Stegmueller) and *constructivist* approaches (Paul Lorenzen, Juergen Mittelstrass), which mostly differ in the degree of emphasis they give to the descriptive or normative perspectives. In all these approaches, the aspect of theory dynamics is dominant.

4. Philosophy, orienting itself on the task of a philosophy of science, stays close to science, and increasingly so even as science is entering in ever closer union with technology and finding new forms of organization. A new approach towards *technology*, as it emerged in the 20th century, is displayed, for instance, in medicine, microelectronics, and laser technology – science is leaving its academic home and is relating its knowledge to the problems of this world more and more often[6] –, a change towards new *organizational forms* through strengthening the extra-university research in the area of basic as well as in the area of applied research – with big centres of sciences such as CERN, EMBL, the Weizmann Institute and the love of large science groups (centres, clusters, networks, alliances).

With these institutional developments, not only has the organizational structure of science changed, but also the *concept of research*. Originally, this concept was closely linked to the researching subject – researchers and not institutions researched – but now the link between research the verb and research the noun is pulling apart. The community of researchers has become Research with a capital "R;" the (re)search for truth, central to the idea of science and at the very bottom of any scientist's self-image of what makes him or her a researcher, has become research as a business operation, an organizable and organized process in which individual scientists, thought to be as interchangeable as individuals in the business world, disappear. The mentioned predilection for core areas, centres, clusters, alliances and networks in research is the embodiment of this change. The change is reinforcing the industrialization of science, but is also weakening science's ability to self-reflect. Self-reflection is a distinctive mark of enlightened science. It is characterized by the right ratio of proximity and distance. This is just as true in institutional terms and, when it is achieved, it constitutes the rationality of institutions, in this case scientific institutions. It is also true where scientific self-reflection is paired with social reflection (in the form of advising politics and society), a link in which modern society can find its true "scientific" character.

There is also a *normative* aspect connected to the idea of self-reflection. Not just epistemological questions, but also aims and objectives are at issue here, and thus questions of orientation, both theoretical and practical. The ethical

6 See J. Mittelstrass, *Leonardo-Welt: Ueber Wissenschaft, Forschung und Verantwortung*, Frankfurt: Suhrkamp 1992, pp. 47–73 ("Zukunft Forschung: Perspektiven der Hochschulforschung in einer Leonardo-Welt" [1990]); H. Nowotny and P. Scott and H. Gibbons, *Re-Thinking Science: Knowledge and the Public in an Age of Uncertainty*, Cambridge etc.: Polity Press 2001, 2007; P. Weingart and M. Carrier and W. Krohn, *Nachrichten aus der Wissensgesellschaft: Analysen zur Veraenderung der Wissenschaft*, Weilerswist: Velbrueck Wissenschaft 2007.

consequences of an increasing scientification of the world, for instance, belong to these. Philosophical foundations – these are not just epistemological, but also practical and ethically relevant foundations, through which science is normatively reconciling itself with itself and society. The fact that also foundational questions such as these have been addressed in the 20th century, together with the significant theoretical breakthroughs and the epistemological debates accompanying them, characterize it as a truly scientific century. At the same time, this character epitomizes demanding requirements which science and philosophy have to satisfy today and in the future.

13 The Scientific Mind. Does Science Make Its Own History?

As card-carrying members of a rational culture that likes to look at itself in the mirror of its scientific capabilities, we often speak of science as if it were a person or a subject, a good acquaintance and neighbour, who goes about the business of knowledge and has important things to say about the future of our world and about how we want to understand ourselves in this world. It solves problems but also creates them and has thus – as they say – problems of credibility. It has moods (what it has just declared to be true it calls false tomorrow); it has quirks (it is infatuated with things that no one else cares about); and it has secrets (its explanations are often as baffling as what they explain). It is prudish when propositioned ("repaying society" is a modern term for this proposition), but it fawns on us when social resources are being distributed. It is like Athena, and we, the scientists of both sexes, work in her garden, increase her prestige and wealth and share with her the fortune of the dissatisfied, whose fortune seems only to consist in truth. What is this, Science? And how realistic is a conception that sees in science the real subject of the search for truth and of the future of rational culture?

13.1 Justified history and factual history

The image of science and the scientific mind just sketched finds its epistemological counterpart in the notion that the history of science displays a *history of progress*; indeed, that science is the actual active subject of the kind of history, which we as scientists would like to see ourselves a part of. Science presents itself here in historical perspective as the gradual transformation of the world into a textbook – true to the enlightenment programme of Immanuel Kant as the emergence of mankind from an intellectual immaturity for which it itself was to blame. For as the grand old man from Koenigsberg put it: "Immaturity is the incapacity to make use of one's understanding without the supervision of another." And we are ourselves to blame for this immaturity, "if its cause lies not in a lack of understanding but rather in a lack of resolve and courage to use it without supervision."[1] And this emergence is precisely what science has written on

1 I. Kant, "Beantwortung der Frage: was ist Aufklaerung? (1784)," *Werke in sechs Baenden*, ed. W. Weischedel, Wiesbaden: Insel 1956–1964, vol. VI, p. 53.

https://doi.org/10.1515/9783110596687-013

its banner, what it exhorts its workers with, and what makes not only the history of science but the history of humankind as well a rational history, a history of progress.

Such a notion, hatched especially in the brains of the philosophers among the scientists, but also firmly anchored in "normal" scientific consciousness, is what philosophers today, this time in the guise of philosophers of science, persistently try to talk us out of. These philosophers view the rationality of scientific practice, previously held sacred, as questionable; they speak in sociological categories about external factors that direct the development of science just as strongly as do internal factors, and even about a "finalization" of science, that is its co-opting by the goals and purposes of an external regimen of needs. Science, to continue the chosen simile, appears as a fickle goddess who has long since left the path of Kantian enlightenment, glances to the right and to the left, is pushed and pulled one way and the other and shares the needs of a society that still expects use and advantage from science, but is miles away from submitting its forms of life to science's idealistic norms. Whoever looks in the history of science for a subject, Science, that reveals itself as the subject of a history of progress oriented toward pure truth and knowledge, will find in the more recent forms of historiography of science only the very human and very humanly-oriented subjects well known from other areas of history.

This will come as a surprise only to those who view science as the handiwork of gods and not of men. But the history of science was for a long time written as if it traced the footsteps of the gods, who stooped to nothing human. This kind of notion is found even in the first origins of modern philosophy and science, for instance, in Francis Bacon's *New Atlantis* (1627), where a happy little island nation, abiding otherwise by biblical traditions, lives off the blessings of technological progress (from desalinization to the construction of various models of perpetual motion), having replaced their political leaders with a regime of scientific experts. Even this early in history, the scientist appears as the new wise man on the way from Faustus to Leonardo and Newton, and the scientific mind appears as heir and successor to God the creator, guaranteeing social happiness. We hear the same from Werner Siemens more than 250 years later: Let us, he said, "hold onto our conviction that the light of the truth that we explore will not lead us astray, and that the power that it confers on mankind cannot degrade us but must raise us to a higher level of existence!"[2] The notion

2 H. von Siemens, "Das naturwissenschaftliche Zeitalter (1886)", in: H. Autrum (ed.), *Von der Naturforschung zur Naturwissenschaft: Vortraege gehalten auf Versammlungen der Gesellschaft Deutscher Naturforscher und Aerzte (1822–1958)*, Berlin: Springer 1987, p. 155.

that scientific progress is irresistible, and the notion tied to it, that the free development of the scientific mind and the increase especially of natural scientific knowledge automatically leads to an increasingly humane society, have molded the self-perception of modern man and the view of the history of science as a history of progress in both theoretical and practical senses.

This view also comes to light in the opinion that the history of science itself might be finite.[3] This is the philosophical thesis of the complete or asymptotic exhaustion of nature, according to which the history of scientific discoveries is either absolutely limited or at some point will enter into an asymptotic approach to what can possibly be known. In this case all scientific problems would be solved; everything there is to discover would be discovered, and everything there is to be explained would be explained. At most a mopping-up operation would be left – calculating a few decimal points and classifying a few cases which tell us nothing really new. Along these lines Denis Diderot assured his readers, "that before one hundred years have passed one will not find three great mathematicians in Europe. That science will come to a dead stop pretty much where a Bernoulli, an Euler, a Maupertuis, a Clairaut, a Fontaine, a d'Alembert and a Lagrange have left it. They have erected the Pillars of Hercules beyond which there is no voyaging."[4] Similar expectations have been voiced with regard to the completion of modern fundamental physics. "It is possible to think of fundamental physics as eventually becoming complete. There is only one universe to investigate, and physics, unlike mathematics, cannot be indefinitely spun out purely by inventions of the mind (...). Some unsolved problems might remain in the domain earlier characterized as organized complexity, but these would become the responsibility of the biophysicist or the astrophysicist. Basic physics would be complete; not only that, it would be manifestly complete, rather like the present state of Euclidian geometry."[5] The same thing is asserted by the thesis of a complete or asymptotic exhaustion of information capacities, according to which it is the reception of the possibilities of scientific information that either are absolutely limited or at some point become an asymptotic approach to an absolute limit of the receipt of information.

Especially in the case of the first thesis, the potential exhaustibility of nature, the omnipotence of the scientific mind would be documented precisely

---

3 On the question of limits of science, see chapter 5.

4 D. Diderot, "De l'interpretation de la nature (1754)," *Œuvres complètes*, vols. I-XX, ed. J. Assézat and M. Tourneux, Paris: Garnier 1875–1877, vol. II, p. 11. See N. Rescher, *Scientific Progress: A Philosophical Essay on the Economics of Research in Natural Science*, Oxford: Blackwell 1978, p. 9.

5 B. A. Bromley *et al.*, *Physics in Perspective I*, Washington D.C.: National Academy of Sciences 1972, p. 80. Quoted by Rescher, *Scientific Progress*, pp. 9–10.

in the closure of scientific knowledge. The end would be the fulfillment of the scientific promise to explain and comprehend the world with nothing left over – to be sure, an unusual way for scientists to become unemployed. What would remain would however at least still be some kind of divine form of existence: Like the eye of God looking with satisfaction upon creation after six days of work, the eye of the scientist could rest on the textbook that explains and comprehends everything. However, even with regard to God's work there are serious doubts whether all was done well.

In the world of science such doubts will certainly have their grounds, though this does not prevent this world from fancying itself to be divinely independent and selfdetermined. Even an approach that does not see the end of all knowledge in the completeness of present or future knowledge but rather writes the history of science as the history of (scientific) problems and their solutions is not free of this fancy. For example, take the following remarks, with which a senior historian of science, Edmund Hoppe, began his *Geschichte der Physik* in 1926: "I call this book history of physics, not history of physicists, not history of natural philosophical views. Of course there are in various periods more or less dominating grand ideas in physics, but they have only been fruitful insofar as they are derived from the research resulting from the study of individual problems. These individual problems must be constitutive of the comprehensive view. But they have a history! And it is the history of these problems that I shall try to present."[6] This is history of science written as a history of problems and problem solutions – sometimes isolated from "general" history, sometimes joined to it, but joined in such a way that "general" history in part acquires the character of a history of progress in problem solving. This is surely not in every case false, but it is, just as surely, an all too harmonious view of human relations, of which the scientific relations are also a part.

Precisely this is the reference point, in various ways, for modern philosophy of science with its philosophy of the history of science. In this philosophy, continuities and discontinuities stand for the historical, contingent character of theory formation, and a theory of theory dynamics stands for the conceptual structure of scientific knowledge. According to Thomas S. Kuhn, who distinguishes between phases of normal science – phases of greater loyalty to a programme or paradigm – and revolutionary phases – phases resulting in the replacement of one scientific paradigm by another –, scientific developments cannot in principle be conceived as a history of progress and thus as a *rational* history caused, among other things, by the asserted incommensurability of con-

6 E. Hoppe, *Geschichte der Physik*, Braunschweig: Vieweg 1926, p. III (my translation).

ceptual structures. Both the resistance of paradigms to anomalies in normal science as well as the methodological indeterminacy of revolutionary phases, that is of changes of paradigms, make scientific developments appear to be partially irrational processes. Even an evaluation of these developments seems not to be possible in a manner independent of paradigms; even this judgment must, according to Kuhn, rely on standards that are themselves part of the developments under review.

Such a notion of the formation of scientific knowledge and its dynamics points in the direction of a thoroughgoing relativism in scientific affairs. Paul Feyerabend's "anything goes," taken methodologically, seems also to determine the relation between scientific and other forms of orientation. But this is contested not only by general common sense but also by scientific development itself, which does not display a capricious vacillation between rationality, or adherence to standards that stabilize rationality, and witchcraft. For this reason also – indeed taking into account the contingent character of many scientific developments – philosophy makes a distinction between *internal* (oriented to standards of rationality) and *external* (political) factors, whereby it is the criterion of methodological reconstructability that divides the process of science into an internal and an external history.[7] Rational reconstructions grasp only a part of scientific practice, namely its theoretical and methodological parts; they need, for historical reasons as well, to be supplemented by historical empirical analyses. These, as it were, bring the human side of science to the fore, while rational reconstructions are directed to their purportedly divine side.

Thus, modern developments in the philosophy of science also reject the radical view that the history of science does not in any way differ from other histories of human relations, not even in its structures of rationality. Those who maintain this not only have science against them (including perhaps their own disciplines); they will also be unable to distinguish between a well-founded scientific development and an unfounded (misdirected) scientific development. Such a distinction goes beyond the positions in the philosophy of science mentioned here, including Karl R. Popper's, which hopes for a gradual approach to the (scientific) truth by means of a consistent pluralism of theories, that is, by generating as many alternative theories as possible under competitive conditions. The distinction requires the justification of *trans-theoretical* criteria that

7 I. Lakatos, "*History of Science and Its Rational Reconstructions*," in: R. C. Buck and R. S. Cohen (eds.), *PSA 1970. In Memory of Rudolf Carnap* (*Proceedings of the 1970 Biennial Meeting. Philosophy of Science Association*), Dordrecht: Reidel 1971 (*Boston Studies in the Philosophy of Science*, vol. VIII), pp. 91–136; also in: I. Lakatos, *Philosophical Papers*, vols. I-II, ed. J. Worrall and G. Currie, Cambridge: Cambridge University Press 1978, pp. 102–138.

cannot be reduced to the conventional commitment to an aim of science (Popper) or to historical demands of consistency. For instance, that every methodology applied to developments in science must itself meet its own standards for scientific theory, or must deliver rational reconstructions that conform to the criteria of the chosen methodology (Imre Lakatos).[8]

Without this kind of distinction, which can also be grasped in the distinction between a *justified history* and a (mere) *factual history*[9] – whereby the concept of a justified history comprehends stages in the rational construction of a science within a factual history of scientific relations – without this, all talk of science and scientific developments will lose its critical thrust or will remain, as in Kuhn's theory of paradigms of scientific truth, part of those scientific relations that it wants to evaluate. In other words: The point is not so much whether internal or external factors govern scientific developments, for instance in so called "revolutionary" phases (Kuhn), or whether science in this sense is the (autonomous) subject of its history or not; rather the point is to come to an understanding of the sense in which scientific developments can be reconstructed by means of concepts like that of a justified history. Where science does proceed on the basis of grounds, it may also be taken as an autonomous subject directing its own history.

Furthermore, in the context of a discussion of science as a subject we should not forget the distinction between science as a particular *form of knowledge production*, namely the production of scientific knowledge, and science as an *institution*, as socially organized knowledge, meaning the "sites" of research: universities, research institutes, etc.. Science is ideally and really always both – methodically enlightened research rationality and institutional reality. This means that the question whether science is the subject of its own history can take either philosophical shape as the question of the autonomous development of its form of rationality, or sociological shape as the question of the autonomous development of its institutional form. In both cases we are dealing with the "independence" of science, and in both cases the development of the form of knowledge and the development of the institutional form of science, are as a rule closely linked. That is why we speak not only of the autonomy of the scien-

---

8 See J. Mittelstrass, "*Die Philosophie der Wissenschaftstheorie: Ueber das Verhaeltnis von Wissenschaftstheorie, Wissenschaftsforschung und Wissenschaftsethik*," *Zeitschrift für allgemeine Wissenschaftstheorie* 19 (1988), pp. 308–327, also in: J. Mittelstrass, *Der Flug der Eule: Von der Vernunft der Wissenschaft und der Aufgabe der Philosophie*, Frankfurt: Suhrkamp 1989, pp. 167–193.

9 See J. Mittelstrass, "*Rationale Rekonstruktion der Wissenschaftsgeschichte*," in: P. Janich (ed.), *Wissenschafttheorie und Wissenschaftsforschung*, Munich: C. H. Beck 1981, pp. 103 ff..

tific mind but also of the autonomy of the university – albeit with decreasing verisimilitude.

13.2 Influencing

If we give credence to the modern skepticism in philosophy of science about the thoroughly rational character of scientific developments, then the path of science must have been determined to a considerable extent – and not just on its institutional side – by external factors. In fact, however, there are only a few really spectacular cases where this is obviously true. Here we are reminded of the sad fate of Giordano Bruno, who was burned at the stake for his doctrine of the infinitude of the world and the plurality and equal value of world systems, and of Galileo Galilei, who was prosecuted for championing Copernican astronomy. But even in these cases, when viewed soberly, the loss of rationality suffered by science due to external intervention was minor. Bruno's "scientific" truth was more the product of an uncontrolled and confused fantasy, and Galileo's scientific truth was at the time still without real physical justification. It is not the case that what science already knew had its influence and development stunted by external factors (here, the Inquisition), but only – and only for a short time – one particular scientific research programme. In Galileo's case his house arrest led him to return to the even more revolutionary programme of developing modern mechanics in the *Discorsi* (1638).

Nonetheless, in the more recent history of science there are also examples of the dominance of the external history of science over its internal history, that is, explicit social and political influences on scientific developments, influences on the acceptance or rejection of scientific theories. Let us take three brief (well known) examples.

1. *Darwin's theory of evolution.* Much of public opinion at the time, especially that of the Church, reacted with horror and outrage to Charles Darwin's theory of the descent of man, which revolutionized biology. This theory, according to some not very subtle accusations, undermines the foundations of civilization and the sanctity of conscience.[10] For this reason the churches and some parts of the intelligentsia attempted to prevent the acceptance of Darwin's theory by the scientific community.

10 See F. M. Wuketits, *Charles Darwin: Der stille Revolutionaer*, Munich and Zurich: Piper 1987, pp. 84 ff..

What is peculiar here – in contrast to the case of Bruno and Galileo, where the objections themselves were presented as scientific arguments – is that the asserted falsehood of the theory is not supposed to be revealed by empirical or other internal scientific grounds. Rather, *ethical*, *moral* and general *social* grounds are adduced (and this applies even today to an otherwise rather antiquated debate in the United States about the status of Darwin's theory). For some the problem is that if the theory is right it will have ethically unacceptable consequences for the picture that humans have of themselves and of the basis of communal life. For this reason, said the social opponents of Darwin, the theory cannot be right. This is well expressed in the nice exclamation attributed to the wife of Bishop Wilberforce on the occasion of Thomas Henry Huxley's lecture on Darwin's theory at a meeting of British natural historians in 1860: she hoped to God that it is not true, and if true that it will not become generally known.[11]

The basic figure of argument is the same as in the classical cases of Bruno and Galileo. The difference lies in the fact that because of the increased institutional independence of science this attempt at influencing the content of science remained without success. Darwin's theory was accepted in spite of external resistance. The "internal" scientific rationality prevailed institutionally as well and without historical delay.

2. *The case of Lyssenko.* This attempt at external influence upon the course of science was more successful than in the case of Darwin, in as much as a (spatiotemporally limited) change in scientific doctrines was implemented by means of political authority. Soviet biology in the 1930s and 40s was obliged by governmental decree to subscribe to a particular theory, namely Trofin Lyssenko's theory of heredity.

The political background of this affair lay in the view that man is purely the product of his environment and that there are accordingly no "constants of human nature" which escape the formative influence of the environment and conditions of life. Based on this approach it could be expected that a transformation of the relations of production will create a new man. The task was to support this theoretical approach – accepted on political grounds – with empirical scientific evidence; and the problem was that traditional genetics countenanced certain traits – Mendel's "factors," precursors of today's genes – which are already fixed at the formation of the germ and to this extent are shielded from the direct influence of the environment. The solution to this problem was, as was to be expected, coarse. It was declared that traditional genetics represented a perspective

11 See H. Luebbe, *Der Lebenssinn der Industriegesellschaft: Ueber die moralische Verfassung der wissenschaftlich-technischen Zivilisation*, Berlin: Springer 1990, pp. 52ff..

distorted by class interests; it was a "bourgeois theory." This notion was supplemented by recourse to pre-Darwinian interpretations of the development of species, namely the assumption of "inheritance of acquired traits." Lyssenko worked out his theory on this basis and reported success in finding empirical support. The theory then became by fiat the official basis of Soviet biology.

The basic pattern in this case resembles the classic examples: we have a *closed* society, a society that rests on a particular state dogma and has institutions to watch over the purity of doctrine. These institutions exercise influence on which theories in science are to be accepted and which are to be rejected, whereby state influence is not restricted merely to rebuffing empirical approaches that are incompatible with the ruling opinion, but also actively puts itself in the place of science and positively favors a particular approach.

3. *Adequacy criteria for IQ-tests*. This example is of a different kind; it deals with political influence where scientific judgment itself is not unequivocal. In this case an inner-scientific space arises that is then occupied by general considerations of an ethical and social nature. Thus we are dealing with the problem of which questions or tasks ought legitimately to be included in an IQ-test. Some criteria of selection are obvious or uncontested. For instance, since the days of Alfred Binet one criterion has been that a task must be solved better by older children than by younger. But criteria of this kind are not sufficient; there is still much free space, which is occupied by other criteria. For instance, ever since the 1930s the criterion has been applied that the aggregate results of an adequate IQ-test should not show any difference in statistical mean between men and women. In other words, if a suggested IQ-test shows the result that the mean values for men and women differ, then the test is inadequate. Here the political principle of equality of the sexes enters into the methods of collecting data.

As opposed to the cases mentioned earlier, this kind of political influence is not in principle reprehensible. What is different here is that it is not the case that an answer given by science is put aside and politically overruled; rather in this case science – that is, the relevant psychological theories themselves – has no answer. But this case is also interesting for another reason. There is no consensus on whether or not other differences are to be treated on the model of sex differences.

Thus, without any claims to completeness, these three cases give some indication of how science under the perspective of its presumed independence from unscientific cooptation "communicates" with society, how internal and external history interact. In the first case internal scientific rationality prevailed institutionally without delay despite social resistance. In the second case (for a time) external factors usurped the place of internal factors, that is, politics (external history) took the place of science (internal history). In the third case internal and external factors join to constitute scientific truth. All three cases are "normal" – and not only as historical incidents. They only confirm the human character of science and the difficulty of keeping a grounded course of scientific knowledge (in the sense of the above-mentioned justified history) free from external cooptation, that affects not only its institutional but also its rational form.

13.3 Basics and applications

Modern developments confirm this normality in a much different and much less dramatic way in the disappearance of a clear-cut distinction between *basic research* and *applied research.* This distinction, which placed basic research as "pure" science in the corner of truth and knowledge, while applied research, as subordinated to external purposes, was left in the corner of industry, has turned out to be too simple. It is ever less adequate to the reality of scientific research, in which discovery, invention, and development have moved ever closer together.[12]

This does not mean that *pure basic research* does not exist anymore. It can be identified as such wherever research is being carried out the results of which show no recognizable practical application. Typical research fields of this kind are high energy physics and cosmology; for instance, the development of a unified theory of all non-gravitational particle interactions or the connection of the theory of elementary particles with the theory of gravity (including the development of models of the so called expanding universe and the prediction of so called "cosmic strings"). In all these cases the notion of application makes no sense, not even a predictive sense (unless one wants to take up *science fiction*). This research is pure basic research.

A somewhat different case is what we might call *application-oriented basic research*, a type of research from which we expect applications in the long run, but not of the kind that could be directly marketed or developed within the normal planning time spans of industrial enterprises. Examples of this would be high temperature superconductivity, synergetics (non-linear thermodynamics) and the foundations of information sciences. The last group includes alternative conceptions such as optical computation, parallel computation, or collective switching systems, whose construction copies neuronal networks. In such

12 See J. Mittelstrass, *Leonardo-Welt: Ueber Wissenschaft, Forschung und Verantwortung*, Frankfurt: Suhrkamp 1992, pp. 60–65.

cases application is intended even though the paths from research to application are unclear and are themselves in need of intensive research.

A third type of research should be distinguished from these two, what we might call *product-oriented research.* This is research which takes place either with a view to particular applications or which promises such applications in the near future. Examples are materials research, environmental research, medical research (for instance AIDS research). In such cases the paths between research and application are short and are concretely part of the research programmes – in contrast to pure and application-oriented basic research.

These three types of research are often mutually supportive in concrete research programmes both in and outside the university. The forms of research interlock and intermingle when focusing on a problem. Pure basic research still exists only in very special research fields. Application-oriented basic research is becoming more and more the norm. The archaic simplicity (sometimes simple-mindedness) in research affairs has become a complex interlocking of interdependent research orientations. This means that the goals of science, in as much as these are expressed by such ideals as truth and knowledge, are more and more joined to the goals of a world that is less inclined to admire than to apply the results of science. From the perspective of science this need not be a limitation. To know what holds the world together is today still an eminent scientific aim; to *hold* the world *together* is surely no less eminent.

The claims to autonomy of science that are expressed in the formula of science as subject of its history are not threatened by the diminishing clarity of the distinction between basic and applied research. After all, neither the Greek mind, to which we owe the idea of science, nor the modern mind, to which we owe the modern world, charged science to stay away from application. However, increasing contact to application means increasing *responsibility.* In addition to the responsibility of science to itself for abiding by standards of rationality such as transparency, reproducibility, truthfulness, and intersubjectivity, there arises a responsibility of scientists for a world that is to an increasing extent the result of their labours through the impact of science and technology. Being a good scientist has become more difficult, not easier. This, too, is a result of the history of science and its embedding in other histories.

13.4 Science as subject?

The view of the philosopher and the philosopher of science is often like that of the owl who begins its flight at dusk, when, as Hegel says, "reality has completed its process of formation and has made itself ready."[13] The owl's view, which is often a historical rather than a systematic view, retraces lines that the winds of history have already blown away, that the bustle of the present has made unrecognizable; it holds fast to what is vanishing (the scientific mind, too, is forgetful); it comprehends, but no longer changes anything. Changes, even in science, are in any case increasingly things that just happen, not things that are strategically or systematically brought about. Already, attempts at a *natural history of science* have been proposed in philosophy of science. Disciplinary developments are described in terms of the theory of evolution with variation and (natural) selection.[14] They are conceived like biological species characterized by various genotypes formed by mutation and constituting a gene pool; variation occurs in theory-variants whose empirical content changes, continuity is anchored in selective factors that determine changes in shape. Scientific developments, philosophically viewed, adapt themselves to natural developments. Furthermore, science has become unwieldy. The notion that it could be the rational subject of a rational history, if only of its own history, is fading away.

This is caused not only by philosophical pictures confusing and fleeting in their multiplicity, but also by the realities of science. Thus, for instance, the concepts of pluralism and democratization have brought confusion to scientific standards, not just in the Sunday sermons of philosophers of science but also in everday scientific life. Paul Feyerabend's already cited slogan "anything goes" – facetiously promulgated against an all too rigid adherence to method that tends to introduce blinders and an inflexible conservatism in scientific affairs – is gaining ground even beyond the more narrow framework of philosophy of science (keyword, once again: theory dynamics). Concepts from the political sphere (democratization, pluralism, liberality) that may well have their legitimacy in the *institutional* side of science are entering its *methodological* side. Those who insist in affairs of method on justification, rigor, and constructive completeness are already suspected of dogmatism and ultimate foundationalism.[15] Fashions are rampant (in philosophy at the moment, they tend to follow

---

13 G. W. F. Hegel, *Grundlinien der Philosophie des Rechts oder Naturrecht und Staatswissenschaft im Grundrisse* (*Vorrede*), *Werke in zwanzig Baenden*, ed. E. Moldenhauer and K. M. Michel, Frankfurt: Suhrkamp 1969 – 1979, vol. III, p. 28.

14 S. Toulmin, *Human Understanding* I (*The Collective Use and Evolution of Concepts*), Oxford: Clarendon Press 1972.

15 See H. Albert, *Traktat ueber kritische Vernunft*, Tuebingen: Mohr 1980, pp. 8 ff.; H. F. Spinner, *Pluralismus als Erkenntnismodell*, Frankfurt: Suhrkamp 1974, and *idem*, *Begruendung, Kritik und Rationalitaet: Zur philosophischen Grundlagenproblematik des Rechtfertigungsmodells der Erkenntnis und der kritizistischen Alternative* I (*Die Entstehung des Erkenntnisproblems im griechi-*

the French); political programmes mix with scientific programmes (e. g. peace research, feminist science) and the programmes themselves are open to such humbug as attempting to overcome modernity. Fundamentalists, relativists, postmodernists and (Feyerabend's) witches are once again knocking on the gates of science. There is no more serviceable bad conscience than one we have been talked into. And just such a conscience is suddenly rampant in science. Or is it really the case that only very few have noticed that overcoming modernity, which includes overcoming the programme of enlightenment at its base, could land us on the tail side of European rationality, namely in a new anti-enlightenment?

Apparently, science, which was once the paradigm of a methodically progressive well-grounded reason and still claims to be so, has become susceptible to programmes and fashions that see reason itself only as one form of orientation among others and equate pluralism (in scientific affairs as well) with relativism. How otherwise can we explain that today anything, just because it is written in the name of science, is also printed, and that scientific rationality, which once seemed to be sexless though not without needs, suddenly discovers its sexuality and becomes either male or female? Pity the Greeks, to whom we owe philosophy and science and in these the discovery of reason. They were truly a sensual little nation and nonetheless in affairs of reason quite capable of unflinching rigor. The world wallows in phantasms and the scientific mind is right in there with it.

As the foregoing examples make clear science has also created a situation in which it is beginning to become unmanageable (also in part as a consequence of its own growth). No one knows any more what science knows; and science itself does not know either. This again is connected with the subjectlessness that is threatening. Scientific change is becoming like so called technological change which transpires without a recognizable subject, so to speak anonymously, invisible to the individual and behind his back. Scientific change becomes incomprehensible to itself, not for philosophical reasons but for contingent reasons that have to do with the loss of standards and of the ability to distinguish between

schen Denken und seine klassische Rechtfertigungsloesung aus dem Geiste des Rechts), Braunschweig: Vieweg 1977; W. Stegmueller, *Probleme und Resultate der Wissenschaftstheorie und Analytischen Philosophie* IV/I (*Personelle Wahrscheinlichkeit und Rationale Entscheidung*), Berlin and Heidelberg and New York: Springer 1973, pp. 25 ff.. In contrast J. Mittelstrass, "Gibt es eine Letztbegruendung?," in: P. Janich (ed.), *Methodische Philosophie: Beitraege zum Begruendungsproblem der exakten Wissenschaften in Auseinandersetzung mit Hugo Dingler*, Mannheim and Vienna and Zurich: Bibliographisches Institut 1984, pp. 12–35, also in: J. Mittelstrass, *Der Flug der Eule*, pp. 281–312.

reasonable developments and unreasonable developments, between a justified history and a factual history.

There is something else also connected with the loss of science as a subject that takes control of its history and thus of its future, demanding rediscovery or restitution. As was emphasized earlier under the headings of reordering of research orientations and the responsibility of the scientist for the results of his labour, the modern world is in great part a product of scientific labour. Without this labour the world would be different (better as the opponents of science mistakenly believe) and without this labour the modern world would have no future. The world listens to science, but – and this is what is peculiar – *science does not speak*, at least not as an (institutional) subject.

This does not mean that we hear nothing at all from the side of science. On the contrary, science has become silent to the extent that scientists have become talkative, as seen in the circumstance that there have never been so many experts among scientists as we find today; they take the floor to speak about anything and everything. The modern world has become an expert's world in the sense that it seeks to meet the multiplicity of its problems with a division of understanding into a multiplicity of problem solving rationalities. Unfortunately this means that the modern world is not governed by Leibnizian monads, in which in questions of knowledge the entire universe is supposed to be reflected, but by the specialist, especially the scientific specialist, in whom almost nothing or only (with apologies to Friedrich Schiller) a divided world is reflected. Those who know more and more about less and less and otherwise nothing at all have landed on the backside of the universality that was once the aspiration of the scientific mind: they seek it in details which for them have become the whole. This corresponds to the self-perception of the modern world that loses its orientation in the fragmentation of its orientational rationalities.

Add to this the fact that the scientist in his role as appointed or self-appointed expert all too often allows himself to be manipulated by parties and other political and economic realities. This is, however, not very helpful for these realities, and it is ruinous for science. The social realities are only played back their own melodies; science is drawn into conflicts of interests and needs that remain themselves songs unsung. The fact that science's subjects, the scientists as experts, are as a rule drawn into spectacular contradictions (at legal hearings etc.) makes its predicament more than clear. It seems to be adapting to just those relations which it ought to help overcome.

In other words, where science gets lost in its scientific subjects, the scientists in the role of appointed or self-appointed experts, where it remains silent as an institution, with the result that it is unrecognizable as an acting, judging, and planning subject (of its history and its future), it fails its most important role in a world that itself is the product of the scientific mind. In this way it becomes an uncritical part of this world, whose standards are power, not knowledge, and interest, not truth. But this is not what was meant earlier by saying that knowing what holds the world together is still an eminent goal of science and holding it together is no less so. The cement that could do this is still orientation towards truth and objectivity whatever the cost, rationality without ifs, ands or buts, responsibility not measured against contingent interests and needs but against the *categorical imperative*. Therefore, science is challenged once more to comprehend itself in an institutional sense more strongly as not merely an investigative subject but also with regard to the just-mentioned standards and orientations as a wise and counselling subject. Only then will science and the scientific mind again be what they are in essence: elements of a rational world.

III Ethical and Institutional Matters

14 The Moral Substance of Science

Science and morals form an ancient topic. Plato and Aristotle had already connected the idea of science with that of morals – in the notion of what the Greeks called a good life, which had to have both a theoretical and a practical form. A theoretical life (βίος θεωρητικός) and a practical life (βίος πρακτικός) go hand in hand. When a practical life lacks a theoretical element it cannot recognize itself (*homo sapiens* without *sapientia*). And when science lacks a moral orientation, that is to say an orientation towards the good life, it remains senseless (a tool without an end). In such cases, a rational culture in which praxis is guided by theoretical considerations, that is to say in which praxis understands itself as being reflected, and in which theory is related both to practice and to life, could not come into being.

This idea of the interrelation of science and morals seems to have got lost along the long roads followed by science and ethics, and along the long road of reflection about science and morals. At least since Max Weber, the idea has taken hold that science is value-free, and that science is formed according to rules differing from those of morals. Conversely, many think that morality has no need of science, in that it is something radically different from scientific rationality. On the side of the sciences, there is also the view that this rationality of the sciences, above all of the exact and empirical sciences, constitutes the whole of rationality. It then follows by definition that any points of view which seek to constrain scientific practice, whether by reference to "practical" or normative considerations, are in fact unauthorised points of view, or indeed ones damaging to science.

But this point of view is itself too radical, for it overlooks the fact that science is not value-free, as the Greeks had pointed out already, and that it rather has a moral substance. This will be taken up in the following under the rubrics "science as idea," "the measure of progress" and "ethos in the sciences."

14.1 Science as idea

As a rule, the concept of scientific rationality refers to a particular *form of knowledge* and its production, that is to say to theories, methods and the special criteria of rationality to which theories and methods are subjected. Among these criteria, whose fulfilment represents a condition on knowledge- and truth-claims, are, for instance, the reproducibility and controllability of scientific results and procedures, the linguistic and conceptual clarity of scientific represen-

https://doi.org/10.1515/9783110596687-014

tations, the intersubjectivity and testability of scientific results and procedures, as well as methods of justification. If such criteria are abrogated, science loses its claim to objectivity and truth, so that science and opinion become indistinguishable. But this is only one meaning of the concept of science, although it is, from the scientific point of view, the most important one.

A second meaning of the concept of science is given by the fact that science is also a *social organization*, that is, the particular social form in which science is realized as a special form of knowledge formation. Here, we speak of science as an *institution*, for instance the university. The formation of science stands under particular socially defined conditions, among which we may include the pedagogical and research responsibilities of the university. Science becomes visible as an institution, even if only symbolically, when one thinks of the invocation of truth and of the spirit which earlier adorned the portals of our universities.

But the concept of science is still not exhausted by this second, institutional meaning. There is a third one extending beyond those of its theoretical and institutional characters. This can be illustrated in connection with the above-mentioned criteria of rationality. These criteria cannot be restricted to purely methodological aspects, especially if, following the sociologist of science Robert K. Merton,[1] we add to them such criteria as disinterestedness, truthfulness, and organized scepticism, that is, the general invocation to criticize. On the contrary, these criteria connect scientific rationality to a *moral form*. With regard to this moral form, science is not only methodically enlightened rationality or a means to differentiate and stabilize the social organization of consumption and the satisfaction of needs, but it is also an *idea* that relates to the second nature of man, i.e. his epistemic or rational nature, or, even more, a *form of life*.

This third meaning, which transcends everything methodological or theoretical and everything institutional, was once the essential meaning of science. Greek philosophy, to which we owe the theory-form of knowledge, spoke expressly of the *bios theoretikos*, the theoretical life, and not of theories that, in the sense familiar today, make up the contents of textbooks. *Theoria*, according to Aristotle, is a general orientation with regard to life; theory in this sense – not in the sense of our textbook concept – is one of the highest forms of practice.[2] The scientific or epistemological subject and the "civic" subject are still one here, and therefore the truth-orientation of science cannot be played off against its social relevance and *vice versa*. With *theoria* as a form of life, truth also be-

---

1 R. K. Merton, *Social Theory and Social Structure*, 2nd edition, New York: Free Press and London: Collier-Macmillan 1968, pp. 604–615.

2 *Eth. Nic.* K7.1177a12ff..

comes a form of life, that is, it belongs not to the methodological but to the moral form, and thus to the idea of science. In this sense both the work of man on his rational nature and truth are moral. How does this express itself in actual scientific and social developments? Is "the idea of science" also actual?

14.2 The measure of progress

Another fact that seems to speak against the suggestion that science has a moral substance, and that scientific rationality orients our life is the progress made by science, and in consequence by technology. For science seems to go where it wills. Furthermore, scientific and technical developments are interdependent. Progress in the one drives progress in the other, and *vice versa*. Progress in science and technology is, at its essence, *immeasurable*, excessive, or to put it differently: if there is an internal measure of science and technology, then it is that they exceed all measure. For measure means definition, or limitation, whereas scientific and technical rationality define themselves precisely through the provisional character of what limits they may have.

Still, that is not all that one can say. If scientific (and technical) progress has no internal measure, a measure which could of course be a moral one, then this means nothing more than that the limits of progress are self-imposed limits, and thus that the measure of progress can only be a self-imposed measure. The idea that the world, that nature itself has limits that cannot be surpassed by the scientific understanding, and that progress also has a measure that delimits it from inside, does not in fact make sense. It is an idea that can be disproved at any time on both historical and systematic grounds. Thus the boundaries of progress do not lie at those points where they are evidently impassable, but rather where they *should* lie, in other words where man decides that he may not proceed further. Self-imposed limits in this sense are *moral* or *ethical* boundaries. The same is true from the point of view of measurement. If there is a measure of progress, then it is not a "natural" measure, but an ethical one. For it assumes an answer to the question concerning which forms of progress man wants, and which he does not, that is to say which forms can be justified by ethical norms and which cannot. At least regarding his ethical nature, man remains the *measure of all things*, just insofar as he resists assimilation by the world – not only in moral and political matters, but also in scientific and technical ones. And this is an idea that has been attached to the concept of science from the very beginning, that is to say from its foundations in Greek thought.

Generally speaking, ethical problems in research and in science, problems concerning the consequences of scientific praxis and progress, are problems of *practical* reason, not of theoretical or technical reason. By this it is meant that in rational or technical cultures, the rational or technical understanding is not in a position to solve the problem of justified progress, or to respond to the demand for a orienting form of knowledge that goes beyond knowledge as a form of mastery. Already Max Weber claimed that "All natural sciences give us answers to the question: What should we do, *if* we want to master life technically? *Whether* we want to master it technically, and whether that indeed makes sense – they leave such questions unanswered, or they assume [the answers] in pursuing their ends."[3] Answering such questions is not the responsibility of science from Weber's point of view. But this just makes the problem concerning a form of practical reason that guides action, thus of a justified progress, all the more troublesome. Science has acknowledged this itself, and has indeed regretted the weakness of practical reason. As Albert Einstein observed in 1948: "The tragedy of modern man lies in the fact that he has created for himself existential conditions that are beyond the capacities given him by his phylogenetic history."[4] Put otherwise, the drives of the subcortical structures are stronger than the cortical control. One might well ask in this situation whether science, in its freedom of research, still bears responsibility for what it does and what it effects?

Freedom and responsibility are difficult concepts not just in the context of science and research. They are among those that everyone has on his lips and some in their hearts as well, even if they do little more with these concepts than to apply them rhetorically. We know that freedom of research or freedom of science is written into the programme of the enlightenment and into many modern constitutions, that research and development serve social purposes, and that responsibility is one of the virtues of a citizen in a democratic society. But it remains difficult to state more precisely what responsible freedom of research or science are, and where they begin and where they end.

In the case of science, the problem begins already with the fact that freedom of research or science means on the one hand freedom of the *scientist* and on the other hand freedom of the *institution* of science. The restriction of the one freedom is often justified by the claims of the other: Since the institution of science is losing its freedom increasingly to the state – so say the scientists – the personal freedom of the scientists must be all the more unrestricted. Since the freedom of the scientist is claimed and exercised without restriction – so say the govern-

3 M. Weber, *Gesammelte Aufsaetze zur Wissenschaftslehre*, ed. J. Winckelmann, 3rd edition, Tuebingen: Mohr 1968, pp. 599–600.

4 A. Einstein, *Ueber den Frieden: Weltordnung oder Weltuntergang?* (ed. O. Nathan and H. Norden), Bern: Lang 1975, p. 494 (my translation).

mental administrators – there must be regulatory influence of the state on institutional affairs. This seems to mean that it is no longer possible to take both the freedom of the scientist and the freedom of the institution of science together. Wherever the one is exercised without restriction, the other must accordingly be limited.

But this surely involves a misunderstanding, one which indeed occurs whenever one fails to make an adequate distinction between freedom and arbitrariness. Often the social good of the freedom of science deteriorates into mere whims on the part of the scientific actors, namely the right to do what they like. Concepts like justification and (social) responsibility seem in the minds of many scientists to belong to the vocabulary of the unfree. But this is mistaken. Freedom, rightly understood, is always *responsible* freedom, otherwise it is arbitrariness. Consequently, both freedoms, the freedom of the scientist and the freedom of the institution go together. Freedom of science understood as a boundless subjective freedom of the scientist is unacceptable from the point of view of science because the old Humboldtian ideal of research in "solitude and freedom" cannot be demarcated effectively enough against misunderstandings of unbounded scientific subjectivity. Even genius, which in scientific affairs is not nearly so common as scientists like to think, does not justify expansion without limit. This holds in science as well.

So much for the concept of freedom of research. The concept of responsibility with regard to this freedom still remains to be discussed. In fact, wherever a claim is made to freedom of research or science, this freedom must be related to structures of responsibility. This leads us then to ethical or moral arguments. Again, the usual distinction between science as a particular form of knowledge formation and science as an institution is not exhaustive. This has been made clear by norms which, serving as criteria of scientific rationality, are above all practical, as opposed to theoretical, in kind. They are aimed at superseding mere subjectivity. Scientific states of affairs are strictly speaking *inter-* or *trans-subjective* states. Not in the sense that scientific subjects disappear, but in that they are distinguished by a morally determined generality of scientific norms such as those mentioned. Those who do not subordinate their work to these norms, which are not purely methodological norms, not only overstep the bounds of scientific rationality, but they also overlook the normative lines that connect scientific work with the life-world. Science has not only a *knowledge*-task but also an *orientation*-task. It has a cultural meaning.

14.3 Ethos in the sciences

In this context, the notion of a scientific ethics is a popular topic of conversation these days. It is supposed to counter the suspicion that not all is well with the ethical bonds that once held between science and society. One hears more and more talk in connection with the sciences about arrogance and immoderation, indeed even about treachery in the ranks. Science's supposedly divine nature has evidently given way to quite human urges.

On the other hand, there is much evidence that the expectations directed towards an "ethics of the sciences" and to its realization are too great. It may even be that the call for such an ethics may lead us in the wrong direction, at least in so far as one thinks of an ethics of the sciences as a special ethics *for scientists.* There cannot be such a thing, for the simple reason that an ethics is always an ethics of the *citizen.* It cannot be divided along social lines, that is to say in a scientific ethics which is the ethics of the scientists, and a non-scientific ethics, which is the standard ethics of society as a whole. And the same holds for morals. There are, strictly speaking, no closed ethical or moral worlds, in each of which a single ethics or set of morals holds sway.

This objection is directed not only at the exaggerated hopes for an ethics of the sciences, but also at the idea that the scientist has more responsibilities than the average citizen. A scientist does of course have a special responsibility, which derives from the essential uncontrollability of scientific knowledge by extra-scientific knowledge, as well as from the dependence of modern society on the special competence of the scientific understanding. However, this special responsibility does not translate into a special ethics. What is needed is rather a better *ethos,* as for instance has long been the case with the socially realized professional ethos of physicians. All rules, all norms which one might like to prescribe to the practice of science in order to strengthen the responsibility of science and of scientific rationality, are superfluous once we have such an ethos of the scientist and once it is in fact observed. Of course that it is in fact often not observed is obvious enough. But that does not mean that an ethics of the sciences has failed, or that it must be improved, but rather that the norms of general, civic ethics were violated, and the ethos of the scientist was violated by base personal motives. There is little more that can be said about the ethics of the sciences, except perhaps that the attention of science as an institution towards the observance of the scientist's ethos should be more strongly enforced in the future.

As an example of this sort of institutional attention we might take a so-called "code of conduct" published in 1998 by the *German Physical Society* (DPG). Here we may read that "Every member is also a member of the community of scientists, and shares in their special responsibility towards coming generations. The members support the development of science. At the same time, they acknowledge and respect the fundamental principle that holds for all science in all countries, namely that of honesty towards oneself and others. The DPG condemns scientific misconduct and disapproves both of fraud in science and of the deliberate misuse of science."[5] Clearly enough, notions deriving from a general civic ethics are being translated onto science and the special circumstances of scientific practice. These rules do not constitute an ethics of the sciences in a distinct sense.

Rules such as these, which science imposes upon itself in order to tie its freedom to some ethical measure, sound like rules of reparation. They hint dimly at some forgotten scientific ethos which conceived of science as an idea and a form of life. Indeed, the ethos of science has today lost much of its effectiveness, and thus also its subjects. However, to the extent to which it has become unrecognizable, it has also lost sight of society and its relation to science. The crisis of confidence that has grasped hold of science is also an ethical crisis, a crisis of a scientific ethos. Thus it is of utmost importance to overcome this crisis that science is itself responsible for.

In this connection, there are three arguments, which on the one hand explain why it has come to a crisis of confidence both with and within the sciences, while at the same time making clear what must be kept in mind in the future.[6] Among the causes of this crisis of confidence is first of all an increasing "scientific incompetence" on the part of society, of which science is of course a part, by which is meant the inability to understand the production of scientific knowledge. A second cause is the "desymbolisation" of science, which has not led to "emancipatory progress," but rather to a loss of "ethical self-consciousness." Third, there has been increased competitive pressure, that is to say an uncritical importation of the market model into the practice of science. Here it is largely a question of reversing this trend whenever possible by appeal to the forms of (social) interaction that are specific to the sciences, and which speak against using an economic paradigm, or indeed using a "professional code" of "institutional procedures."

These are indeed essential factors in questions of confidence and ethics, and yet, in the final analysis, it is a matter of most importance to bring back a scientific ethos to scientific consciousness. We understand under the notion of an ethos an orientation towards largely implicit, and implicitly observed rules,

5 "Verhaltenskodex für DPG-Mitglieder," *Physikalische Blaetter* 54 (1998), No. 5, p. 398.

6 C. F. Gethmann, "Die Krise des Wissenschaftsethos: Wissenschaftsethische Ueberlegungen," in: *Ethos der Forschung / Ethics of Research (Ringberg-Symposium Oktober 1999)*, Munich: Max-Planck-Gesellschaft 2000, pp. 38–39.

which are conceived as holding self-evidently both for individual and social actions. Whether we conceive of these rules as the simple rules of conduct to which one usually holds (rules of etiquette), or whether they are rules to be evaluated morally or ethically, such as maxims – in both cases it is a matter of implicit knowledge. And this knowledge demands being followed practically rather than being theoretically mastered.

The connection between an ethos, morals and ethics would then be the following. Ethics is a critical theory of morals, which is above all concerned with regulating institutional morals that are often in conflict with one another. That is to say with regulating socially implanted systems of rules of action and goals by evaluating them and deciding among them by providing the arguments that permit decisions. These arguments must in consequence be generally valid, and so the corresponding ethics must itself be *universal*. This means in turn that it makes universal claims of validity, and that it must be in a position to ground these claims. Immanuel Kant's ethics provides an example of such a universal ethics. An ethos is, on the other hand, a part of morality, and thus of a universal morality when the latter is characterized by a universal ethics. Here, an ethos relates to a universal conception of ethics, that is to say it "represents" the latter's claims to validity, or indeed it realizes them.

And just this is the case with science. For science is the expression of universal claims to validity, and this both in the sense of being a special form of knowledge formation, that is to say of the scientific formation of knowledge, as well as in the sense of being a scientific ethos, which is also the moral form of science. The orientation towards truth typical of the one of these follows the orientation towards truthfulness of the second. That is to say, quite simply, that *truth* determines the scientific form of knowledge, whereas *truthfulness* determines the moral form of science, which as a result belongs to the form of life of the scientist, to his ethos.

Our task for the future is thus to make these connections explicit in the practice of science, and to ensure that we act in accordance with that explicit knowledge. For if this cannot be achieved, then the crisis of confidence into which science has fallen – deservedly and undeservedly – will continue. This will in turn threaten not only the foundations of science, but also the foundations of rational cultures in general, that is to say of modern society. The question concerning the ethics and the ethos of science is therefore not merely a question concerning the future of science, but also one concerning the future of our society, concerning that of our culture.

15 Science and Culture

That “thinking is one of the greatest joys of humankind,” i.e. thinking precisely in its scientific form, is one of the things Bertold Brecht has his Galileo tell us.[1] And since thinking is culture, science too is culture. Has the modern world, which has science as one of its conditions of existence, forgotten this today? On this question four theses calculated to display the cultural nature of science and its constitutive character for the modern world.

1. *Science is not an external aspect of rational cultures. On the contrary, it is precisely in science that the rational nature of the (modern) world and of man is realized. The result is a Leonardo world whose future lies in the pursuit of a scientific or research imperative.*

On the long road from the discovery of the scientific mind by the ancient Greeks to our times, modern cultures have learned to define their rationality primarily in terms of scientific rationality, and also in terms of technical rationality. These cultures are therefore the product of a rational history. They are, moreover, the historical product of the *epistemic* and *technical* nature of man – his searching and finding, planning and building, thinking and doing. The concept of epistemic nature here means that the formation of knowledge is not external to man. Rather, it constitutes the medium in which man *orients* himself – in everyday and scientific situations. Where there is no knowledge, man loses his orientation or becomes the tool of foreign, unfathomable knowledge. Orientations are here understood as including not only knowledgeable ways of dealing with objects and their relations, but also their interpretation and explanatory appropriation. In scientific knowledge this appropriation acquires its true rational form. What Aristotle formulates in the proposition that all men by nature strive for knowledge[2] is, if understood in this way, the grounding of rationality in the epistemic nature of man.

Man’s technical nature corresponds to epistemic nature on another level. When man creates tools for himself, and relies on tools for his way of life, he adds, as it were, artificial organs to the other organs through which he interacts with the world. Technology not only supplies what is needed for the “completion” of man’s nature; it also develops its own laws, acquires its own rationality, and creates a new nature. The historical product of this constructive power, along with the epistemic power mentioned, is the modern world. *Homo sapiens*

---

1 B. Brecht, *Leben des Galilei*, in: B. Brecht, *Versuche 12–15*, Frankfurt: Suhrkamp 1977, p. 296.
2 *Met.* A1.980a21.

https://doi.org/10.1515/9783110596687-015

and *homo faber* realize the peculiar double aspects in which the rationality of man and his world develops. This also means that in cultures that are, in their existence, a product of scientific and technical rationalities, it is not just any of man's abilities or "talents" are realized, but those that make him a rational subject and his world a rational world. The modern world as the reality of scientific and technical cultures, as they are represented by modern societies, is not a contingent, but rather a logical result of man's developed nature.

Thus, we come to understand that our world is the work of man. "Natural" worlds exist only on the margins of this world, and they are becoming ever fewer and ever weaker. This again is not an incidental but a logical result of the developed nature of man. Science today is everywhere, as is technology. Wherever we go in this world, the modern mind has already been there: grounded on science and technological know-how it produces, builds, administers, and destroys. Let us call this world the *Leonardo world* after Leonardo da Vinci, the great Renaissance engineer, artist, philosopher, and scientist. It is a world in which man no longer moves merely as a discoverer, as a stranger in a strange land – let us call this a *Columbus world* – but rather a world in which man is constantly confronted with his own work, a world that in the hands of the scientific mind is becoming ever more an *artifact*, fragile like nature but ever less natural.

The Leonardo world is, in this sense, an artificial world, and there is only a constantly shrinking natural world beyond its boundaries to correspond to it. The Leonardo world has become boundless. This means, in turn, that science, the constructor of this world, is drawn ever deeper into its own world. Man confronts himself in his own works and has become a part of his own works. One example of this is, of course, the flawed concept of a fixed external nature. Ever since the hominids became man, this assumed nature has been becoming culture. There has never been an "untouched" nature in the environment of man. Clearing, burning, hunting, ploughing furrows, redirecting watercourses, digging up the earth in search of minerals, and producing waste – in short, consuming, changing and substituting natural resources – man has always turned his environment into a *cultured nature*. He intervenes in the evolution of his environment, changes it, directs it and makes it useful to him. The environment of man becomes the work of man.

The second example is the concept of society: Modern societies have become dependent on science to such an extent that every change in the system of science (and technology) immediately touches their foundations. This applies both to the existence of scientific progress and to the possibility that it might fail to materialize. This is particularly clear in the sectors of food, health, and energy. Without science there is nothing that can be done; without new energies invested in science the world will run out of steam. The same applies to the sectors of information, transportation, and everything else that characterizes this world as a Leonardo World.

A final example. What holds for modern society as a whole also holds for modern man. Science has begun to see man himself as a potential new Leonardo world and to claim him. We have gradually got used to the growing know-how of reproductive medicine and we are getting used to the notion that the nature of man (see genetic engineering) can just as well be changed as can the physical and social worlds. Just like the physical and social worlds, man is becoming more and more an artifact. We are taking our evolution into our own scientific hands in a consistent manner unthought of by earlier societies.

All three examples show that the Leonardo world is growing and that science, linked with technology, is its real engine. This applies, as the examples also show, not only to the sunny side but also to the dark side of the Leonardo world. Hopes are attached to the notion that all problems caused by science may also have scientific solutions, and worries are attached to the notion that the man of the Leonardo world might be the loser in this development. Today, those who believe the Leonardo world already to be lost are clearly gaining an audience and (social) influence, and hold science, above all, responsible for its fate. The *Zeitgeist* paints in dark colours and science is a favourite subject.

Certainly no one will deny that science not only solves problems but also creates them. The danger to the biosphere, for instance, is not least the result of successful scientific (and technological) rationality (see fluorocarbons). However, we would make it all too easy on ourselves, if we believed or persuaded others to believe that we could enter calmer waters simply by giving a command that would curb science and technology. It is true that we are faced with a dilemma of progress (supported by science) and its consequences, of scientific change and technological consequences; it is false, on the other hand, to believe that this dilemma could be resolved by restricting scientific (and technological) activities. Thus, it would be a simple mistake to believe that the problems of a Leonardo world, even those which are consequences of its scientific nature, could be mastered with less science (and less technology). The opposite is the case: marked decline in scientific research and technological development would bring the Leonardo world to a state of incapacity to act or react. The problems of a Leonardo world are not static, they do not stop moving when we stop, or when science stops.

Thus, we have implicitly formulated a *scientific imperative*, a research commandment. While following this commandment will not deliver us once and for all from the dilemma of research and its (unwanted) consequences, ignoring it will lead the Leonardo world ever deeper into self-caused and other problems.

This same point is implicit in the idea that the world is a product of the epistemic and technical nature of man, that science is the nature of a Leonardo world.

2. *Science has an epistemic, an institutional, and a moral form. As such it is itself an integral part of culture – of culture understood as the system of all human forms of work and life.*

Science in a Leonardo world increasingly puts itself in the place of nature, but is it therefore already *culture?* For everyday consciousness, science consists of theories and methods that are only understood by the scientific consciousness itself. Science is seen as a largely un-understandable context of research and action carried out by a small group within society, the scientific community, which deals with what makes this society a modern society and its world a Leonardo world. Such a society recognizes its reflection in the products of science but not in science itself; and as a rule, it defines its culture by means other than science.

However, the concept of science is not a simple concept nor is that of culture. First of all, science is a particular *form of knowledge* and its production, it is secondly an institution, and thirdly an "ideal" or, even more, a "form of life."[3]

Today, we are far removed from the reality of this third meaning. By way of false alternatives like science-for-its-own-sake or science-as-a-pure-productive-factor in the life of industrial societies, we have become estranged from the notion that science is a way of living. Accordingly, our universities no longer transport forms of life with which students and teachers could identify as in the sense of the old *universitas magistrorum et scholarium*. The same holds for culture (if one restricts the term to the so-called culture industry, segregated from the labour form of society), that is, to those things we do or visit after work (for instance theatres, museums, and concerts). These are culture, too, but not all of it. Culture means two things: It is the system of human forms of labour and life and it is a subsystem of this general culture distinguished from such areas as technology, economy, and politics. This is not contradicted by talk about technical culture, economic culture and political culture; rather it marks the transition from the concept of culture in a particular sense (which includes the culture industry) to the concept of culture in a general sense; what is meant is again culture as the system of all human forms of labour and life.

One part of the general culture is made up of the sciences. Not only because in them the epistemic side of human nature is expressed, but also because, as the conceptual explication of science was supposed to make clear, they belong

3 See chapter 14, for more detail.

to the labour and life form of a modern society. Taking science out of its connection to culture would thus itself be a phenomenon of deculturation; the losers would be science (which would then lose its moral form) and general culture (which would thus lose a good part of its rational form). In other words, science and culture do not in any important sense oppose one another. On the contrary, science is of its very nature culture, and without science the culture of a Leonardo world would lose its genuine, its rational nature.

3. *Culture moves in the medium of symbols and interpretations. Scientific theories, too, are interpretations. They make neither themselves nor the world that they describe definite.*

A cultural world is always a symbolically mediated world, an interpreted world. It is not only in the transformation of the world into an artifact (Leonardo world) that an appropriation of the world transpires, but in our interpretations of it as well. In this sense the Leonardo world is also a *Leibniz world*, that is – alluding to the concept of perspectivism in the philosophy of Gottfried Wilhelm Leibniz – a world with which man is connected by his interpretations. According to Leibniz the unity of the world is given objectively. This shows itself, however, in a plurality of different perspectives from the differing points of view (*points de vue*) of the *monads*. According to this conception each of the monads reflects the universe in a different way. This applies not only to culture in the narrow sense, for instance, literary culture, but also to science as an integral part of rational culture. Even (scientific) theories are interpretations. It would be a serious misunderstanding to attempt – perhaps with the distinction between explanation and understanding – to draw a clear-cut boundary between scientific knowledge and cultural opinion. Even if there is unanimity in science about theoretical mechanisms, these can still nonetheless be interpreted divergently.

Thus, various interpretations of quantum mechanics differ on the question of whether the quantum mechanical indeterminacy is the expression of an essential contingency or rather is merely the result of imprecise knowledge of the realized framing conditions, as is, for instance, postulated by interpretations by means of hidden variables. Another example is the interpretation of the electromagnetic field as a state of a mechanical ether in the mechanistic tradition of the 19th century. Departing from this interpretation, Albert Einstein conceived of this field as an independent magnitude. Both are different (possible) interpretations of the same Maxwellian theory of electrodynamics. Finally, it is disputable whether a relational theory of space, according to which space represents merely a relation among objects and does not itself exist beside the objects or outside them, is really adequate for the *general theory of relativity* – as Einstein himself believed. Depending on how one translates classical relationalism into the concepts of relativity theory, one receives different answers to the question. At the moment at least, it is impossible definitely to privilege one particular one of these translations.

In other words: One and the same theoretical approach can be differently interpreted; interpretations in these scientific cases, too, are not unequivocal. On the contrary, they display characteristic uncertainties that cannot be completely removed even by a rational reconstruction of the basic principles underlying a theory. The interpretation of quantum theory is not essentially different in this regard from an interpretation (say) of Immanuel Kant's theory of space and time.

What holds for culture in general also holds for science. Interpretations (the field in which culture moves) here in their theoretical form do not stop at science but also determine this form of knowledge formation. The knowledge of the perspective character of the world, whose (scientific) founder was Leibniz, comprehends its theories as well and, at the same time, also makes something clear that, in an epistemological sense, is the essence of all truth, namely the appropriation of the object by means of its *representations*.[4] Interpretations, too, are in this sense representations which conform to principles and not only to the things they represent. These principles can, in turn, not simply be derived from the things; they are not part of the world, but rather parts, perspective parts, of our reflection upon and research into the world. In a Leibniz world, which is only the other side of the Leonardo world, the counterpart of knowledge is not an "absolute" world, and the counterpart of the world is not "absolute" knowledge. Rather, things are as we see them and represent them – through our life world experience and through our scientific theories. There is one exception: Just as we cannot put things in the place of experience and theories, so too, experience and theories cannot take the place of things. Theories do not make the world disappear, they let it be represented. That is culture in both the narrow and broad sense.

4. *The quarrel about the "two cultures" is historically understandable but systematically an unnecessary quarrel. There is only one scientific culture, and this is an expression of the epistemic nature of man and thus also of the Leonardo world, to whose cultural form all sciences belong.*

One expression of the insecurity in conceiving of science as culture is the discussion about the dualism of the natural sciences and the humanities that fol-

4 See J. Mittelstrass, "Philosophie in einer Leibniz-Welt," in: I. Marchlewitz and A. Heinekamp (eds.), *Leibniz' Auseinandersetzung mit Vorgaengern und Zeitgenossen*, Stuttgart: Steiner 1990 (*Studia Leibnitiana Supplementa*, vol. XXVII), pp. 1–17, also in: J. Mittelstrass, *Leibniz und Kant: Erkenntnistheoretische Studien*, Berlin and Boston: Walter de Gruyter 2011, pp. 85–107.

lowed in the wake of Charles Percy Snow's concept of two cultures. The humanities, on Snow's initial thesis, represent culture, the natural sciences do not. Whatever difficulties one might have with Snow's ideas for other reasons, he did in fact hit the right nerve. The humanities tend to the view, "that the traditional culture is the whole of 'culture', as though the natural order did not exist. As though the exploration of the natural order was of no interest either in its own value or its consequences. As though the scientific edifice of the physical world was not, in its intellectual depth, complexity and articulation, the most beautiful and wonderful collective work of the mind of man."[5] In Snow's provocative example: reading Shakespeare is culture, but knowing the second law of thermodynamics is apparently not culture.[6]

This line of thought is quite mistaken, not only in the eyes of Snow. It is prevalent because the Leonardo world, from the perspective of culture and the systematics of science, is a particularized world; that is, because the boundaries of professions, of theories, of experiences, of perceptions have become boundaries of individual and disciplinary worlds. And here, too, Snow was right: "In our society (that is, advanced western society) we have lost even the pretence of a common culture."[7] We must do something about this, for instance by making clear that science is not only culture but above and beyond this, that it is the essence of rational culture.

In science (including the humanities) all rational human faculties come together as in a mirror. Science is not only knowledge in the hands of scientists, it is research and teaching, explaining and discovering, inventing and understanding, playing and riddle-solving, ordering and systematizing, building and planning, construction and criticism, and much more. Science is the world of *homo sapiens* and *homo faber*. That is what was meant earlier by speaking of the foundation of the modern world in the epistemic nature of man.

Furthermore, science in progress, like Achilles in another field, represents the eternal youth of the scientific mind, to which a Leonardo world owes not only its origin but also its nature. What allows it to grow old (its Achilles' heel, to continue the metaphor) is only (scientific) dogmatism and a world that no longer trusts in, or has become tired of, its innovative power and seeks to return to myths. And this applies again not only to the natural sciences (including medicine) but also to the social sciences and humanities. The social sciences are not irrevocably wedded to the present, which they seek empirically

5 C. P. Snow, *The Two Cultures and a Second Look: An Expanded Version of the Two Cultures and the Scientific Revolution*, 2nd edition, Cambridge: Cambridge University Press 1964, p. 14.

6 *The Two Cultures and a Second Look*, p. 15.

7 *The Two Cultures and a Second Look*, p. 60.

to comprehend, nor are the humanities wedded to the past, which they seek to interpret. In view of the problems of a Leonardo world, they, too, must think of something new. Of them, too, we should be able to say (as we do of the natural sciences) that they have the future in their bones. If they offer us only the fleeting present and the mild afterglow of the past, then the distance between the natural sciences (which in ever novel ways occupy the new) and the humanities and social sciences will become ever larger. The losers would not only be the humanities and social sciences themselves but also the Leonardo world, to whose cultural form all sciences belong.

16 Naturalness and Directing Human Evolution

The European tradition of anthropology has always distinguished between the biological and the cultural nature of man, in other words between what is natural to him in a physical and biological sense, and what pertains to him culturally, what his cultural essence is. This, however, does not mean that both essences, the physical and the cultural, fall apart, and that therefore, as René Descartes for example holds, man disintegrates into two essences. On the other hand, by establishing the distinction between the biological and the cultural nature of man, problems arise concerning the concept of *naturalness* applied to man. Is this concept only applicable to his biological nature or essence, or does his naturalness consist precisely in that it is expressed by both natures or essences, that is to say by their unity?

In fact, man is a natural being, who can live only as a cultural being. *Descriptively*, within the context of biological systematics, mankind is a sub-species of the species *homo sapiens*, namely *homo sapiens sapiens*, and is the only recent member of the genus *homo*. But this definition includes only the empirico-physical side of man, not that which makes up the nature or essence of humanity *ascriptively*, namely its form of self-description and (not conclusively established) self-determination. The latter was described classically as the *animal rationale*, a being endowed with and determined by reason, or as a being lying between animal and God. More recent anthropologists (after Friedrich Nietzsche) capture this notion in the concept of a "*nicht festgestelltes*," i.e. a not-yet-determined, being (both biologically and culturally). It would be a category error to interpret our actions and thoughts as the products of natural processes, whereby even the act of interpreting becomes part of nature, a "natural fact." But we fall into a new form of naiveté if we oppose this interpretation with a claim that scientifically discovered facts have no influence, or at least ought to have no influence, on the self-determination of man. Thus it is a matter of adopting a scientifically informed and philosophically considered position, one which is beyond mere *biologism* and *culturalism*, which, in other words, is beyond an absolute distinction between biological and cultural explanations, and which refers to both the lives and the laws that shape our lives. Such a position should neither reduce man to (pure) nature, nor to the (absolute) spirit he aspires to be.

In the following, some considerations about what is called in philosophy and theology the *conditio humana* and what role the concept of naturalness could play in this context. Then the question will be raised in what respect the relation between naturalness and the power of directing evolution, particu-

https://doi.org/10.1515/9783110596687-016

larly man's own evolution, create serious anthropological and ethical problems. And this with regard to a future which is not only human, but also humane.[1]

16.1 The natural and the artificial

Modern philosophical anthropology takes its point of departure from two opposing conceptions: that attributed to Max Scheler and that of Helmut Plessner. According to Scheler, "man" is the "X that can behave in a world-open manner to an unlimited extent."[2] According to Plessner, "man" is characterised by an "eccentric positionality,"[3] whereby his eccentric existence, which does not possess a fixed centre, is described as the unity of mediated immediacy and natural artificiality. Accordingly, Plessner formulates three fundamental laws of anthropology: (1) the law of natural artificiality, (2) the law of mediated immediacy, and (3) the law of the utopian standpoint.[4] Similarly, Arnold Gehlen states the thesis that man is by nature a cultural being,[5] and in doing so, his cultural achievements are seen as compensation for missing organs and "man" is defined as a creature of defect (*Maengelwesen*).[6] Common to all these approaches is that man has a particular nature and that it is an essential element of this nature to work on it.

Stipulations of a similar kind can also be found in the history of philosophical anthropology. Thus man is called the creature without an archetype by the

1 For some aspects of what follows see J. Mittelstrass, "The Anthropocentric Revolution and Our Common Future," in: W.-K. Raff *et al.* (eds.), *New Pharmacological Approaches to Reproductive Health and Healthy Ageing* (*Symposium on the Occasion of the 80th Birthday of Professor Egon Diczfalusy*), Berlin and Heidelberg and New York: Springer 2001 (*Ernst Schering Research Foundation. Workshop Supplement*, vol. 8), pp. 57–67.

2 M. Scheler, *Die Stellung des Menschen im Kosmos*, Darmstadt: Reichl 1927, p. 49.

3 H. Plessner, *Die Stufen des Organischen und der Mensch: Einleitung in die philosophische Anthropologie*, Berlin and Leipzig: Walter de Gruyter 1928, pp. 362ff..

4 H. Plessner, *Die Stufen des Organischen und der Mensch*, pp. 309–346. See K. Lorenz, *Einfuehrung in die philosophische Anthropologie*, Darmstadt: Wissenschaftliche Buchgesellschaft 1990, pp. 102–103.

5 A. Gehlen, *Anthropologische Forschung: Zur Selbstbegegnung und Selbstentdeckung des Menschen*, Reinbek: Rowohlt 1961, p. 78.

6 A. Gehlen, *Der Mensch: Seine Natur und seine Stellung in der Welt* (1940), 9th edition, Wiesbaden: Akademische Verlagsgesellschaft Athenaion 1972, p. 37. In a biological definition, "cultural" is applied "to traits that are learned by any process of nongenetic transmission, whether by imprinting, conditioning, observation, imitation, or as a result of direct teaching" (L. L. Cavalli-Sforza and M.W. Feldman, *Cultural Transmission and Evolution: A Quantitative Approach*, Princeton: Princeton University Press 1981, p. 7).

Italian Renaissance philosopher Giovanni Pico della Mirandola: he himself, according to the will of his creator, is to determine the "form," that is, the cultural form in which he wishes to live.[7] According to Immanuel Kant, the question "What is man?" can only be answered if we already have answers to the questions, "What can I know?," "What ought I to do?" and "What may I hope?"[8] The attempt to determine "1. the source of human knowledge, 2. the extent of the possible and profitable use of knowledge, and finally 3. the limits of reason,"[9] is itself an anthropological research programme, and on the background of the critical philosophy of Kant it is an open research programme that defines man according to what he can achieve in theory and practice. For Friedrich Nietzsche, finally, man is the not yet determined animal,[10] and thus science too is seen as the expression of the human endeavour "to determine himself."[11] Furthermore, one of the reasons for the difficulty of saying what man is lies in the fact that man is the (only) creature that possesses a reflective relation to itself, that man, as Martin Heidegger says, is the creature "that in its being relates understandingly to its being,"[12] or that it is "concerned in its being with this being itself."[13] This opens up a broad horizon for an answer to the question, what a human being, what his nature is. The only thing that is clear is what, with regard to the essential openness of man, can be called the *anthropologically basic situation.*

It is equally clear that a differentiation between that which has *become*, which has occurred without any influence of man, the natural, and the *made*, which has been created or shaped by man, the artificial, is not easy to draw, and due to new possibilities of manipulation, not just of nature generally, but also of the (biological) nature of man, it is getting even more and more difficult. The differentiation between the natural and the artificial, however, is still the essential differentiation on which our orientations are based. Even though we know that man has taken a hand in much of what we consider natural, for instance climate or the flora, and that creating the artificial is natural to man,

---

7 G. Pico della Mirandola, *De hominis dignitate. Heptaplus. De ente et uno, e scritti vari*, ed. E. Garin, Florence: Vallecchi 1942, p. 106.

8 I. Kant, *Logik* A 25, *Werke in sechs Baenden*, ed. W. Weischedel, Darmstadt: Wissenschaftliche Buchgesellschaft 1958, vol. III, p. 448.

9 *Ibid.*

10 F. Nietzsche, *Jenseits von Gut und Boese* (1886), in: F. Nietzsche, *Werke: Kritische Gesamtausgabe*, ed. G. Colli and M. Montinari, vol. VI/2, Berlin: Walter de Gruyter 1968, p. 79.

11 F. Nietzsche, *Nachgelassene Fragmente Fruehjahr 1881 bis Sommer 1882*, *Werke*, vol. V/2 (1973), p. 533.

12 M. Heidegger, *Sein und Zeit* (1927), 14th edition, Tuebingen: Niemeyer 1977, pp. 52–53.

13 *Sein und Zeit*, p. 12.

we still use this distinction as orientation. After all, what would a world look like in which this distinction, the distinction between the natural and the artificial, could not be drawn? And how could it be possible to achieve a self-understanding that forgoes this distinction?

Philosophical views that reduce the one to the other, in which everything either turns into that which has become, or into the made, illustrate that such ideas nonetheless play a role in thinking about man and his world. For Arthur Schopenhauer, for instance, in his fiction of a contemplative "clear world-eye,"[14] everything is purely given, unchangeable by human wants and actions, while, by contrast, for Johann Gottlieb Fichte, everything, also the natural, is constituted by an absolute I or self.[15] In the one case (Schopenhauer) everything would be nature, in the other case (Fichte), everything would be spirit.

It is not just our natural intuitions, our way of dealing with the world and ourselves, that speaks against such radicalizations, so does a more detailed analysis of the implicit conceptualization of that which has become, i.e. the natural, and the made, i.e. the artificial. In actual fact, we are always dealing with, in the terminology of Plessner, a *natural artificiality* (as opposed to something seemingly created out of nothing, thus being artifical) and an *artificial naturalness* (as opposed to something seemingly given without intervention, thus being natural). Here, a distinction made by Dieter Birnbacher is helpful to understand the concept of naturalness, namely that between a genetic and a qualitative naturalness, or a genetic and a qualitative artificiality, respectively: "In the *genetic* sense, 'natural' and 'artificial' make a claim about *the manner in which a thing has been created*, in the *qualitative* sense, they make a claim about its current *characteristics* and *appearance*. 'Natural' in the genetic sense is that which has a natural origin, 'natural' in the qualitative sense is what cannot be distinguished from what is found in nature."[16] This distinction in turn may be connected to the scholastic distinction between a *natura naturans* and a *natura naturata:* "The genetic concept of naturalness relates to the aspect of *natura naturans*, that of a creative nature, the qualitative concept relates to the aspect of *natura naturata*, that of nature as nature having the properties it does."[17] This also illustrates that already tradition

---

14 A. Schopenhauer, *Die Welt als Wille und Vorstellung* I § 36, *Saemtliche Werke*, ed. A. Huebscher, vol. II, Mannheim: Brockhaus 1988, p. 219.

15 These examples are to be found in D. Birnbacher's writings, on whose detailed analyses of the concept of naturalness will be drawn in what follows (*Natuerlichkeit*, Berlin and New York: Walter de Gruyter 2006, p. 3).

16 D. Birnbacher, *Natuerlichkeit*, p. 8 (my translation).

17 *Ibid.*

has noticed the dialectical nature of the concept of naturalness, the reciprocal determination of natural artificiality and artifical naturalness.

16.2 Homo faber

Today developments in biological and medical knowledge place man in the unique position of being able to change not only nature in a general sense, but also his own nature, namely to intervene ever more powerfully not only in evolution in general but even in his own. And he is on the brink of changing the measures with which he previously described and regulated his situation, that is to say, the human condition.

While we have known since Charles Darwin that man, not only from the point of view of philosophy and culture, but also biologically, has no fixed essence, he is nevertheless subject to evolutionary changes, even though this is imperceptible to the individual and only recognizable to science over long periods of time. And it has become clear in the light of the new biology that man can intervene in these changes himself – an ability to deliberately change his own genetic constitution, and that of his progeny. In fact, the *conditio humana* itself is changing: in the sense that now even man's biological foundations are at his disposal. This creates a completely new and momentous situation in the domain of anthropology as well as in the domain of ethics – although the idea of determining our own nature is nothing completely new.

In 1488, Giovanni Pico della Mirandola wrote the following about God's intentions towards man: "We gave you neither a fixed dwelling, Adam, nor a particular appearance, nor any special talent, in order that you might have and own the dwelling, the appearance and the talents that you desire for yourself. (...) We made you neither heavenly nor earthly, neither mortal nor immortal, so that you might form yourself as your own, worthy, free and creative sculptor."[18] One hundred years later (1596) Johannes Kepler writes in the dedication letter of his *Mysterium cosmographicum:* "We perceive how God, like one of our own architects, approached the task of constructing the universe with order and pattern, and laid out the individual parts accordingly, as if it were not art which imitated nature, but as if God himself had looked to the mode of building of man who was to be."[19]

18 *De hominis dignitate* (...), pp. 105–106.

19 *Prodromus dissertationum cosmographicarum continens Mysterium cosmographicum*, in: *Gesammelte Werke*, ed. W. v. Dyck and M. Caspar and F. Hammer, Munich: Beck'sche Verlagsbuch-

What Pico della Mirandola and Kepler still affirm in a pious and expressive language is nothing other than the extension of the concept of man as *homo sapiens* to include that of *homo faber*, both with regard to himself and to his world. Pico della Mirandola's characterization of man as "his own sculptor" again corresponds to Nietzsche's and Plessner's definition of man as the not-yet-determined animal, or indeed to Plessner's characterization of man by means of his eccentric positionality (which is juxtaposed to the undistanced centricity of the animal). Similarly, Kepler's characterization of a *homo faber* competing with God paradigmatically corresponds to the modern notion of scientifically supported technical cultures, in which man creates and encounters – both in and by means of his productions – not only the world, but indeed himself.

Is man his own work, in the way that the (modern) world is his work? Certainly not in the sense that man is an artifact that created itself. For even in his role as *homo faber*, and independently of the complementary definitions of his natural artificiality and his artificial naturalness, man remains bound to what has been called the *conditio humana*, and what is meant by the work-like character of man is above all his self-determining ("cultural") essence, not his biological essence. Nonetheless, such distinctions, which are also boundaries, are beginning to give – not only in an epistemological and anthropological perspective as explained here. Against the background of modern scientific and technical developments, the possibility has raised its head that along with the rational nature of man (that which makes him *homo sapiens*) we might change not only his external (physical and social) nature but also his internal (biological) nature. Is his naturalness at risk? Is it at all possible to define this in any detail in a context that is not epistemological or anthropological? And how about the ethical question?

16.3 The ethical question

The recourse to naturalness, which is epistemologically and anthropologically mostly unproblematic, is, however, problematic ethically, in particular, when ethical conclusions are drawn from definitions of naturalness of man. In such cases, what counts as natural lays claim to moral validity, for instance in Hans Jonas, who declares the natural as the highest norm and views any inter-

handlung 1937 ff., vol. I, p. 6 (1596), vol. VIII, p. 17 (1620). Translation by A. M. Duncan, *Johannes Kepler. Mysterium Cosmographicum / The Secret of the Universe* (Introduction and Commentary by E. J. Aiton, with a Preface by I. Bernard Cohen), New York: Abaris Books 1981, pp. 53/55.

vention into natural processes which might be of ethical relevance as an offence against "naturally" given norms, as something against "the strategy of nature."[20] According to Jonas, this is also and, indeed, in particular, valid with respect to the naturalness of man.

Such views immediately provoke the charge of a naturalistic fallacy, in so far as, apparently, an inference is made from an "is" (a given naturalness) to an "ought" (naturalness as a principle or norm).[21] Strictly speaking, however, this charge may only be voiced or, rather, upheld, when an actual *inference* is made from an "is" to an "ought." If instead it is merely used as a point of departure – as compassion is used in Schopenhauer, or the will to power in Nietzsche, understood as a natural inclination of man –, the emphasis shifts towards the plausibility of that approach itself, in this case, towards the previously described "dual nature" of man, expressed in the concepts of natural artificiality and artificial naturalness. Thus it would be an anthropological premise, from which certain conclusions are drawn in an ethical context. In any case it is a *material* approach that causes the problems, if any; the fact that something in particular, namely the natural – in other cases of ethical reasoning it might be conceptions of the good, the just, or the rational[22] – is meant to play the role of a norm or justificatory authority.

The question then is again what may or should be called "natural." Clearly, nature as a whole cannot be meant with this, but also a recourse to man as natural being would not go to the heart of the matter, as illustrated by the complementary concepts of natural artificiality and artificial naturalness. After all, ethics (and morality, of which it is the theory) is always the manner in which man deals with his natural inclinations and needs, thus cultivating them.[23] Kant even declares this the "essential purpose of humanity," that is, as the purpose in the realization of which the true nature of man finds its expression: "Whoever subordinates his person to his inclinations, acts against the essential purpose of humanity, since as a freely acting being he should not be bound by his inclinations,

---

20 H. Jonas, "Lasst uns einen Menschen klonieren: Von der Eugenik zur Gentechnologie," in: H. Jonas, *Technik, Medizin und Ethik: Zur Praxis des Prinzips Verantwortung*, Frankfurt: Insel-Verlag 1985, p. 179.

21 See D. Birnbacher, who considers in great detail the most important arguments against naturalness as a principle or norm (*Natuerlichkeit*, pp. 17 ff.).

22 See O. Schwemmer, "Ethik," in: J. Mittelstrass (ed.), *Enzyklopaedie Philosophie und Wissenschaftstheorie*, 2nd edition, vol. II, Stuttgart and Weimar: J. B. Metzler 2005, pp. 404–411.

23 Again, see D. Birnbacher, *Natuerlichkeit*, pp. 49–50.

but should instead determine them through freedom, as when he is free, he must have a rule, but this rule is the essential purpose of humanity."[24]

Connected to this purpose in Kant is the concept of *dignity*, making reference to the "dignity of a rational being,"[25] in more recent discussions the concept of a *species ethics*. This concept – and thus a "moralization" of human nature – is used by Juergen Habermas against interventions in the integrity of the human species, for instance using the means of reproductive medicine.[26] Thus the natural foundations are at issue, and, in that sense, again what is essential to human nature. If we also count the cultural nature of man as human nature, the fact that man is by nature a cultural being, in other words, that the definitions of natural artificiality and artificial naturalness are again applicable, interventions in his biological nature would change his entire nature – in a manner that possibly cannot be calculated or controlled. Thus the request for a species ethics.

In the Kantian tradition, such an ethics is only conceivable if it is, at the same time, a version of a *rational ethics*, that is, of an ethics that has its universal basis in a formal principle formulated in accordance with the *categorical imperative*, or else biological classifications or categories would take the place of ethical categories. But this means that an ethics of human nature that may be called a species ethics is not, if properly understood, an ethics of a particular kind, that might possibly be subject to the charge of a naturalistic fallacy, but an implication of a rational ethics, with which the principle of human dignity, which, speaking with Kant, expresses "the dignity of a rational being," is applied to the entire human species.

16.4 Concluding remark

Will man put at his own disposal all the "parts" that make up his essence – body, soul and reason? Has he become master of his own nature in a sense which would have been unimaginable even for Pico della Mirandola or Kepler? We must accustom ourselves to the fact that this disposal of man over himself

---

24 I. Kant, *Eine Vorlesung über Ethik*, ed. G. Gerhardt, Frankfurt: Fischer 1990, p. 135 (my translation).

25 I. Kant, *Grundlegung zur Metaphysik der Sitten*, *Werke*, vol. IV, p. 67.

26 J. Habermas, *Die Zukunft der menschlichen Natur: Auf dem Weg zu einer liberalen Eugenik?*, Frankfurt: Suhrkamp 2001, p. 27. The discussion in D. Birnbacher, *Natuerlichkeit*, pp. 169 ff., and M. Kaufmann/L. Sosoe (eds.), *Gattungsethik: Schutz fuer das Menschengeschlecht?*, Frankfurt: Lang 2005.

will increase, driven as it is by scientific and technical development. But we must at the same time preserve, in opposition to this development, those indispensable things which are experienced in love and in happiness, in sickness and in death, and in which, despite the threat of the triumph of *homo faber* over *homo sapiens*, an essential part of our humanity is contained. Might this be what Pico della Mirandola meant when he had God say to man that the latter was created neither heavenly nor earthly, neither mortal nor immortal?

Movements exist today that do not want to stop there. So-called "posthumanism" or "transhumanism"[27] is endorsing a perfectioning of man, made possible by technological and medical advances, as well as the overcoming of the limitations of the species man which have been taken as natural till now. The question here is not merely whether this is playing God[28] or whether a new Pandora's Box is opened, but also, as to what species man might be considered to belong if, as envisioned, he would have left his own species. After all, things such as the experiences of contingency, of neediness, and of ageing[29] are at issue here, which until now had been considered constitutive of the human species. But independently of that, this example equally illustrates the difficulties generally involved in a definition of how human nature is to be understood.[30] But it is also clear, on the other hand, that it is not just the perspective of biological evolution, thus a *descriptive* perspective, but also the perspective of cultural evolution, thus an *ascriptive* perspective, that will play a role.

This may be illustrated in yet a different manner. God's order to man to subdue the Earth certainly did not include the order to subdue himself, neither in the categories of master and servant, nor with respect to his essence, which is reflected, for instance, in the previously mentioned experiences of contingency and neediness. Wherever man attempts to modify his own essence, his own nature, he is at risk of losing his very nature, the nature that makes him human. Natural artificiality and artificial naturalness would lose their balance. Man

27 See L. M. Silver, *Remaking Eden: Cloning and Beyond in a Brave New World*, New York: Avon Books 1997, London: Weidenfeld & Nicolson 1998; N. Bostrom, "In Defence of Posthuman Dignity," *Bioethics* 19 (2005), pp. 202–214.

28 See M. Midgley, "Biotechnology and Monstrosity: Why We Should Pay Attention to the 'Yuk Factor'," *Hastings Center Report* 30 (2000), No. 5, pp. 7–15; L. R. Kass, "The Wisdom of Repugnance," *The New Republic* 216 (1997), No. 22, pp. 17–26; *idem*, *Life, Liberty and the Defense of Dignity: The Challenge for Bioethics*, San Francisco: Encounter Books 2002, 2004.

29 See also C. F. Gethmann *et al.*, *Gesundheit nach Maß? Eine transdisziplinaere Studie zu den Grundlagen eines dauerhaften Gesundheitssystems*, Berlin: Akademie Verlag 2004, pp. 10–23.

30 See on this N. Roughley, "Was heisst 'menschliche Natur'? Begriffliche Differenzierungen und normative Ansatzpunkte," in: K. Bayertz (ed.), *Die menschliche Natur: Welchen und wieviel Wert hat sie?*, Paderborn: mentis 2005, pp. 133–156.

would assimilate with his creation; he would return to a paradigm of machinery, which has already unsettled thoughts and feelings once before, in early modernity. For after man there would not be man, but a product (of man), setting about to take the place of man. The *conditio humana* would become a *conditio technica*, the species man would have ceased to be itself. It would have crossed species borders. But this also means that man, in a certain sense, cannot be optimized, at least not insofar as with such an optimization he would step out of his own nature – however difficult it might be to define that in any detail.

17 Through a Glass Darkly. On the Enigmatic Nature of Science

The veil of ignorance that, in the opinion of many, today hides science from society, and the public from science, and which in consequence often hinders communication about science, might be more easily lifted if the following question is answered: How comprehensible is science, and how comprehensible can and should science be?

Even the most energetic attempts at making science comprehensible, and at winning society for its enterprise, cannot get around the fact that the scientific understanding remains in many cases a mystery for the unscientific understanding. One cannot simply conjure the comprehensibility of what science knows into being. Difficult scientific subjects cannot simply be translated in all their aspects into colloquial language and concepts. He who nonetheless perseveres in the attempt is often disappointed, and this disappointment cannot be laid at the feet of science. Science is in a well-defined sense unavoidably incomprehensible. It is concerned with things that are not understandable to the layman either directly or indirectly, unless of course he his prepared to transform himself over the course of a long apprenticeship into a scientist. And science speaks a language that only science itself can properly understand. This mutual untranslatability belongs to its essence, and is indeed intimately connected to the responsibility of science. Simply and a bit exaggeratedly, put: science loses its scientific character when it is made understandable, and few scientists can be prepared to make such a sacrifice; conversely, everyday experience becomes incomprehensible when it is rendered scientifically.

Fortunately, this is not the last word on the matter. The reason is that what has been said should caution against the suggestion that only arrogance and elitism hinder scientists from making themselves comprehensible. Indeed, the relation of science and society concerns much more than the popularization of what science knows. It is a matter of assuring ourselves, and of justifying the scientific essence of our world, up to and including the structures of our everyday, so-called "life-world." This makes understanding science, or surmounting the inability to understand science as part of a larger, cooperative enterprise, a crucial task. This inability is shared by both scientific and unscientific minds; it can be refuted with three short observations.

Science is problem-solving with other methods. If one thinks of science only as a dialogue of an Absolute Spirit with itself, or as a kind of hermetic world in the head, one has failed to understand both the real essence of science and its task. The essence and task of science consist in the overcoming of problems. These

https://doi.org/10.1515/9783110596687-017

problems are often self-posed problems, but they are nevertheless similar to the problems that the scientific mind poses to itself. This begins (in Presocratic thought) with attempts at explaining the rainbow and perception, proceeds by way of astronomical models and theories of time, and it ends (for the moment) with the mapping of the brain and the decoding of the gene. But whoever knows the problems can understand the solutions. And so both for the scientific and unscientific minds it is ultimately a matter of making scientific problems comprehensible. For how can one understand answers if one does not understand the questions they are supposed to answer?

Science is discovery beyond the frontiers of the evident. Scientific problems are often, if not always, solved by means of discoveries. For instance, in the case of the constitution of matter this solution was provided by the discovery of the atomic nucleus (1909 by Ernest Rutherford). But this is also the case with everyday problems and our dealings with them in everyday experience, for instance when we are looking for the right path, or for the proper spice for a soup. Not only the problem-structures of science and the life-world are similar in this regard (expectations are not fulfilled, experience is disrupted), but the structures of their solutions are similar as well.

Science is a highly stylized form of pre-scientific forms of knowledge. Science has been characterized from its Greek origins on by its *theoretical forms*. One such form, which is also the form of our textbooks, is for example that of the proof. This is indeed the trademark of science – and yet it finds its partner in the everyday world, namely in the form of argumentative communication. Induction, which is the route from the particular to the general, and deduction, the route from the general to the particular, are not just instruments of the sciences. But this means in turn that the world of science and the life-world are connected to each other by means of argumentative structures and structures of action. It is only that in the one world, that of science, stricter rules hold than in the other one, the life-world. These rules mark the path from the experiential form of knowledge to the theoretical form of the latter, a path which as a result does not lead us away from our common world, but rather deeper into this very same one by means of explananation and justification.

Let this suffice as an answer to the question about how comprehensible science can or ought to be. A last remark concerning this "ought."

The path of a science that seeks to make itself comprehensible is beset by risks, and by the enemies of understanding, among whom number quite a few scientists. For if it is true that comprehensibility in the world of science cannot be merely willed into being, that difficult scientific subjects cannot be arbitrarily simplified, and that scientific terminology cannot completely be translated into everyday language, still there is a converse possibility, namely that of needlessly complicating matters in the name of science. This possibility is abetted in many areas by a *jargon of incomprehensibility* that does not advance science, but preserves it from presumptuous attempts to render it understandable. And this holds true just as much of the language in which science is expressed as it does of the theories in which it is represented. They are often like the emperor's new clothes, above all in the social sciences and the humanities, which are always under a special pressure to justify themselves, so that the flight into terminological fancy and esoteric language becomes particularly enticing. Simplicity appears as the enemy of one's own claims to significance. But this means that science here legitimates itself with the credentials of its incomprehensibility (for after all, no one understands the language of modern cosmology, and it is most certainly a science). One speaks the language of Absolute Spirit, which reveals itself only to the initiated, to which one would of course like to belong.

Unfortunately such cases are by no means rare. But to think this way is to subvert the efforts of serious scientific and unscientific minds to orient themselves in a common and comprehensible world. For this reason, and in the light of the far-reaching "speechlessness" between science and society, a *critique of science* that is informed by the latter and practised with care is as important a task as the efforts of everyday understanding to comprehend the world of the scientific mind.

18 Quality Assessment in Higher Education Institutions – from the Perspective of Those Assessed

Nowadays the universities are spared nothing. It is not only the problem of how to finance themselves that is becoming bigger and that the student/faculty ratio is becoming more and more unbearable: universities also suffer from (fashionable) ideas tied to the "spirit of the times." First, in the 1960s and 1970s, they had to cope with the fact that all university relations had to be assessed in terms of sociology; then in terms of didactics and now in terms of evaluation. "I am evaluated, therefore I am," could be the motto guiding the way in which the system of higher education looks at itself nowadays. What concerns me here is that it is such a difficult self-assessment and possibly such a one-sided one.

The author is an academic – even though he belongs to the special guild of the philosophers – and he speaks about science (*Wissenschaft* in the broad sense) from the perspective of science, not from a meta-perspective, at any rate not from the perspective of an administering and examining intellect. Therefore, one should not expect either deep methodological insights as to how quality assessment should function in science, or detailed knowledge about the state of the art of quality assessment, or, for that matter, any practical examples of it. The author is attentive observer of relevant efforts, but he is, first of all, someone who is cognicent that academic practice contains its own self-confirmation, that it does not get trapped in a false paradigm, and that all that we do – also following the key word quality assessment – promotes science and does not impede it in its essential capacity. In this sense – from the perspective of the assessed academic and the assessed academic institution – here are six short remarks.

18.1 First remark

Quality assessment procedures for higher education institutions in Europe were developed first in the mid-1980s. Most European countries have systems of quality assessment or quality assurance at their disposal.[1] The reason behind this de-

1 For an overall view on this development see St. Schwarz-Hahn and D. F. Westerhijden (eds.), *Accreditation and Evaluation in the Higher Education Area*, Dordrecht and Boston and London: Kluwer 2004; G. Haug, "Quality and European Programme Design in Higher Education," *European Journal of Education* 38 (2003), pp. 229–240; R. Lewis, *Recent Developments in National,*

https://doi.org/10.1515/9783110596687-018

velopment has been, and is, to give more autonomy to higher education institutions and to ask for efficient accountability. This is a noble aim, but with the chosen methods also one easily missed. The danger is that by attempting to subject a practice (the academic practice) to standardized criteria, this practice may lose its essential capacity. In the case of science, this essence is in the discovery of what is *new*. This may come in many ways, well-known and new. Therefore, they are not easy to lay down from the start, or to restrict under rules to be followed and controlled, for example in terms of quality. This has to do amongst other things with the fact that in science – as also in many other social areas – *people* are the essential factor, not the routines they follow (in which people are viewed interchangeable commodities). At the centre of research, especially that which is successful, is the researcher, not the research system, be it assessed or not.

By the way, in terms of financing and funding this means that the question is not that of the necessary expenditure for manufacturing a product whose desired function one knows already, but rather that of financing an activity from which science itself expects new insights ("blue sky" research). Society does benefit also from this activity – though in most cases not in the short term – in the form of knowledge and insights which eventually can be used also for practical purposes, the purposes of economy and industry. This makes it necessary to act cautiously with the tool of quality assessment, behind which there are also always an economic interest and associated consequences.

18.2 Second remark

Quality cannot be defined independently of given circumstances – aims, goals, methods, subjects. There is no general definition of quality, and no model that could stand for all areas where quality is at stake. This applies also to research and teaching, which is why quality assessment in (institutions of) higher education is still, in a way, an art without a master. This again means that quality assessment is, on the one hand, a (mostly imperfect) tool that is supposed to solve problems of academic self-perception, and on the other a problem in itself. On this point, the fact of a constantly growing industry of assessment and evaluation should not deceive us. Where everybody assesses everybody else – and

---

Regional and International Quality Assurance Systems (OECD/Norway Forum on Trade in Educational Services: Managing the Internationalisation of Post-Secondary Education, 3–4 November 2003 Trondheim, Norway), URL:http://www.flyspesialisten.no/vfs_trd/ufd/2%20History.pdf

we are moving in this direction – the blade of criticism becomes blunt; what in former times used to serve well-defined aims of optimizing research and teaching becomes an end in itself. We know this from science policy studies, namely from (empirical) research about the way in which, and under what institutional conditions, research is carried out. But this does not make research better, rather it considers it as an object which can be examined like any other object.

18.3 Third remark

Quality in higher education follows particular profiles and scientific practices. It is no absolute measure, it is more a relative one. Relative not in the sense of indifference to what is being considered, but in the sense of taking into consideration all the particular circumstances, aims and goals of a research and teaching institution at any one time. Research is not always the same research, and teaching and study are not always the same teaching and study. Research is organized differently at CERN and in different universities; and teaching and study are organized differently at Oxford, Goettingen, or Padua. General standards may be the same (research, teaching and study at the highest level): but the reality is different each time (roughly depending on the importance of the institution or on the relation of differentiation and standardization in research and teaching) and must also be judged differently. In the end, total uniformity would also mean the end of the need for quality assessment.

In each case, the different stakeholders in the system of higher education expect different things. This is also true with respect to quality. For some, quality, for example, pertains to practices that are economically optimal. For others, for example in the area of basic research, this view can just impair quality, in this case a research-based education. All this will depend again on the choice of indicators, and these will never be the same with regard to individual institutions and to the whole system.

18.4 Fourth remark

If science does not itself know what scientific quality means in research and teaching, i.e., if science itself cannot distinguish between quality and lack of quality, then quality assessment cannot do so either. For, quality assessment pursues its activity according to generally accepted scientific standards. Actually, there is the threat of a circle here if no clarity exists concerning the concept and essence of research quality; for quality assessment in science (dealing with scientific research and teaching) operates according to standards which it derives from scientific practice itself. This is all right if it is clear what research quality is (here the circle is dissolved or else it does not even arise), but it is not all right when there is no clarity (and the circle triumphs), for then quality perishes. The situation resembles that of didactics in higher education. In order to fulfill its scientific tasks, it also must correspond to those scientific standards which it seeks to influence. That frequently this is not the case, determines the obvious weakness of didactics in university teaching.[2]

One more point: quality assessment in research comes either too late – science has already passed judgment (for example in the case of a Nobel prize) – or else it fails to recognize, in its predominant orientation towards routine, the essential people who do not let themselves be constricted by any routine (either in research or teaching). One could also call this the misery of quality assurance, which nowadays is written large everywhere: (with any luck) it ensures what is normal, and together with it, also what is average, but not great achievements, which at best it manages to acknowledge. For great achievements there are no rules, and certainly not those that quality assessment favours or seeks to establish.

And on this point one more observation. Academic achievement of a high calibre and scientific excellence are once again only possible in an environment that is conducive to achievement, that stimulates and furthers academic achievement through academic achievement itself. Although mediocre conditions do not necessarily exclude a high level of achievement, nor occasional feats of excellence, these will remain the exception. Mediocre conditions are rather a programme for academic mediocrity – true to the old university saying, that second-rate people hire third-rate people, ensuring in other words that the tree next door does not grow too tall. Put in another way: there must be a lot of "academic quality" in one place, if academic excellence is to be developed. And this quality is not to be found in isolated fields or disciplinary islands which one occupies alone, but in an academic and scientific context which is defined by quality and excellence. In this sense one may also distinguish excellence from genius: while genius (which by the way is nowhere near as common as academics often think) may grow anywhere, even on academically barren ground, this is not the case with excellence. Academic excellence is awakened by an environment of quality and excellence, it does not develop itself – and if it does sometimes, then very infrequently, as in the case of genius.

---

2 See J. Mittelstrass, "Vom Elend der Hochschuldidaktik," in: J. Mittelstrass, *Die Haeuser des Wissens: Wissenschaftstheoretische Studien*, Frankfurt: Suhrkamp 1998, pp. 213–231.

18.5 Fifth remark

Things look a little different if science is considered not so much with regard to its research activity, but to its teaching activity. There is not just poor research – research which does not deserve this name and something that science itself quickly acknowledges as such – but also poor teaching, which often remains unrecognized even in science: because the scientific mind sees itself more in terms of its research activity than its teaching. Here again it is not simply a matter of declaring the whole affair a didactic matter and then making pedagogues masters of all teaching activity. In the end, a bad teacher is not always the person who does not match conventional didactic standards, but rather the one who has nothing to say, namely the one who does not feel up to productive research. Instead excellent researchers are mostly seen as great teachers, no matter how their teaching is judged from a standard didactic perspective. Albert Einstein is one such example.

What does this mean here? Quality in science has more to do with scientific achievement than with its dissemination, and this includes the context of teaching, because brilliant research inspires more than didactic presentation. And here, the common criteria of assessment fail. As already noted, these are more oriented towards average achievement than towards excellence. They do not explain excellent achievement, and they cannot programme it either. Science goes wherever it wants. And it is no longer to be seen when the judgement takes place.

18.6 Sixth remark

Quality assessment in the academic world has difficulties not only with the essence of scientific research and teaching, but also in the force-field of science and society. Already defined as an "impossible mission,"[3] it must "make the open self-criticism of evaluation palatable to the university community between the inner life of science and society, and dissuade the external world from great interventionist hopes which it connects with assessment, so that assessment becomes feasible."[4] It is mainly the political mind which expects the fulfilment of

3 U. Teichler, "Hochschulevaluation und Hochschulmanagement im internationalen Vergleich – einige Thesen," in: M. Roebbecke and D. Simon (eds.), *Qualitaetsfoerderung durch Evaluation? Ziele, Aufgaben und Verfahren von Forschungsbewertungen im Wandel*, Berlin: Wissenschaftszentrum Berlin fuer Sozialforschung 1999, p. 111 (my translation).
4 *Ibid.*

its aims, in this case of political-scientific aims, from "technical" considerations, and it is the scientific mind itself which replaces here trust with distrust, adjusting to mainstream thought. After all, the dominance of quality assessment has already led to a situation where success is copied and innovation is avoided, because it is always unsure of realization. What is feared, not unjustly, is a new "arithmetic" of academic relations, which we already know in curricular development.

But there is yet another danger. The scientific system is inventive: already, it is no longer simply a matter of assessing the institutions of higher learning, but also of assessing the assessments. In the end, nothing escapes the examining mind, not even the mind itself. And this can be continued: assessing the assessment of assessments, and so on. Is there a better example of a *reductio ad absurdum?* Overassessment is already now the motto of the hour, and woe to us if institutions continue to grow in this way. We are then all constantly assessing and being assessed. Science, which once was at the core of a scientific life, becomes secondary, an assessed secondary subject.

One more such example. By the terms evaluation and quality assessment, we usually mean two things that we should actually distinguish carefully: that which in the context of evaluation and quality assessment figures as *accreditation*, and that which signifies *ranking* (or scoring). The former seeks to set out the minimal requirements of an educational system. The latter tries to compare teaching institutions in terms of efficiency and to bring them into a hierarchical system. The former is necessary and commendable, but often boring, the latter is unusual and demanding, but normally problematic and speculative in its realization. The former keeps to given standards, the latter invents them. In one case, it is the organizing mind which is active; in the other it dreams. And again it is only science itself which could ensure clarity here – not assessed science, but rather science that researches.

18.7 Conclusion

In order not to misunderstand ourselves: in a world where science has become a large concern, where nobody – not even in one's own narrow circle – can see anymore all that science knows and does not know, methods of self-assessment are imperative. But we should protect ourselves from the possibility that it becomes an industry of its own and that, in the end, those who have a say are not the productive researchers, but rather those who are not themselves at the forefront of research, and instead – and that is the power of the image – watch from the stage and try to pass judgement.

19 The Joy and Woe of Scientific Policy Advice

Scientific policy advice seeks the golden mean between political abstinence and its own politicization. In the scientifically imprinted cultures of modern society, we deal with scientific knowledge transfer in political issues, where a lot depends on both, namely science and politics, preserving their own independent character: science its fundamental *orientation towards (scientific) knowledge*, politics (e.g. governments) its substantial *orientation towards action.* This is where the actual problems of scientific policy advice today are to be found: Is such a mutual independence possible when at the same time effectiveness is required or is the relationship of science and policy to another essentially an illusion? Policy advice in idle motion? But is it really running idle? On both sides, the conditions are often unclear. Some comments to the point.[1]

The relation between science and society has never been simple. Science was expected to deliver all that society needed – minds as diverse as Plato and Francis Bacon may be mentioned here – , or, at other times, science caused societal trouble – Nicholas Copernicus and Charles Darwin can be named as examples. And even if society was receptive, assumed scientific unworldliness and chronic incomprehensibility (or opacity) marred the delicate relationship. It has not changed until today. Science should be of this world and simple – and this is rarely the case. The picture of an ivory tower does not only represent the autonomy of science – it could (should?) once again be so understood considering the increasing covetousness of economy and politics – , but also lack of contact on both sides. So what does scientific policy advice mean?

It means that under the heading of "advice" (or "counsel") science and society ought to effect a sort of rapprochement, that science should be applicable without becoming political – a hard job. After all, scientific reason follows the concepts of truth and justification, whereas political reason in its orientation towards action adheres to the concept of power and practical efficacy. This means: science in the role of political advisor from the point of view of science is seen as an attempt to reform the world after all; from the point of view of politics, science is seen as means to its own ends. So what do we have: The attempt of science to gain political influence quickly unmasks itself as an illusion, and on the side of politics the attempt to make science subservient forces science to think in

1 The following comments continue on an earlier statement: "Athena oder Aschenputtel? Der wissenschaftliche Verstand unter Zwaengen der Politikberatung," *Gegenworte: Hefte für den Disput ueber Wissen*, ed. Berlin-Brandenburgische Akademie der Wissenschaften, Berlin: Akademie Verlag 2012, No. 27, pp. 15 – 17.

https://doi.org/10.1515/9783110596687-019

unfamiliar categories. Disillusionment seems to be programmed, again on both sides.

But should this be the last word? Does scientific policy advice remain at idle speed, i.e., does it not propel anything? If that were the case, it would mean that although for a large part the modern world turns out to be an achievement of science, science itself has no influence on its world, and that the modern world is indeed aware of its dependence on science, but nevertheless attempts to direct science for its own purposes, keeping its relation to science arbitrary. Neither of these conditions is desirable. The actual problem of the relation between science and society, science and politics, lies in the *institutionalization* of scientific policy advice.

Indeed, many frolic in the field of (also institutionalized) policy advice nowadays: companies specializing in this form of counselling, professional bodies, firms and business associations, usually acting as lobbyists which means: advice not for the benefit of politics (e.g. governments), but for their own profit. Advice is then marketed as promotion of public good, yet actually serves as means to the benefit of its own. Here, science is no exception. It also becomes a lobbyist when acting primarily on its own behalf, for its own promotion. Borderlines become quite vague. By offering its know-how in matters which politics cannot assess – knowledge for orientation and problem solving – science emphasizes its irreplaceability, thereby, in the case of a state-funded institution, reinforcing its right for state subsidies. This obligation is towards the scientific system in general, in some cases also towards scientific institutions which have a contractually prescribed advisory relationship with the state, which is recompensed by funds for the institution. A recent case in point is the *Leopoldina*, the oldest academy of science in Germany, where its promotion to a national academy is linked with a statutory commitment for scientific policy advice.

A debt to be discharged – a concept which in earlier times was meant to remind science that it owed a service to the society which supported it, namely to supply application-relevant and therefore practicable knowledge – develops into a contract which even then has to be met by science, when it pursues its own scientific aims differing from those which are expected by politics or even stipulated – possibly with a vague promise as to later benefits. This could become *symbolic* idle motion: science somehow fulfils its obligations by summarizing – relevant or not – what it knows, but that is all. In specific cases this could even serve politics very well, in that no obvious obligation for action arises from the information science offers and the role of politics as head of procedures is brought to light more distinctly. It is in the logic of advising or advice, also of scientific advice, that one cannot determine commitments oneself.

Thus, the relationship between science and policy remains non-binding which usually is of little import when both science and politics meet as separate worlds, but this leads to considerable frustration on the part of science, hiring itself out to no effect. According to its own laws, science is used to the fact that knowledge, scientific knowledge, is binding. This is where knowledge differs from opinion. Science has to learn that its knowledge is handled like any other commodity, and at times has to cede to a dominant (political) opinion.

In the context of advice, what does this mean for a relationship between science and politics with clear methods and functions? On the one hand, an eligible codex of policy advice from science and institutionally organized advice on the other. If this organization is based on a contract between scientific institutions (with respective institutional funding) and political institutions, the problems just mentioned will occur. Here, not the scientific process itself determines the course, but the ordering party will, or will at least steer part of the scientific work into certain political channels. Science as an institution becomes itself part of the political process.

The principles which should be followed by scientific policy advice like independent advisory councils for science and research are in brief:[2] (1) The mandator or any third party must not interfere with or exercise an influence on the ongoing process of advice, neither methodologically nor in substance. (2) The responsibilities of the mandator and of the advisory board have to be strictly demarcated from each other. (3) The selection of the scientists engaged in scientific policy advice have to follow criteria which are justified by objective reasons only. (4) The participation in the field of scientific policy advice as a rule has to be honorary. (5) The results of scientific policy advice must be made public, any divergences have to be accounted for. If these guiding principles are ignored in the process of scientific policy advice, the danger, as already mentioned, arises that science itself becomes part of the political process.

In Anglo-Saxon countries it functions differently. It is not a scientific institution, institute or academy that acts as *science advisor*, but a scientist. Becoming a science advisor by a political decision, he passes – and this is the decisive factor – from the house of science into the house of politics. He employs his scientific knowledge in order to give policy advice and can take this stance without having (or being able) to speak in the name of science – he would not be legitimized by science (as an institution) to do so. For example, in Ireland a *science advisor* is expected "to provide high-level advice on scientific issues of concern to Govern-

2 See *Leitlinien Politikberatung*, ed. Berlin-Brandenburgische Akademie der Wissenschaften, Berlin: Akademie Verlag 2008, pp. 25–38.

ment across the spectrum of disciplines," "to provide scientific input to the development and review of the Government's *Strategy for Science*,"[3] in the USA "to advise the President and others within the Executive Office of the President on the impacts of science and technology on domestic and international affairs," "to lead an inter-agency effort to develop and implement sound science and technology policies and budgets," and "to evaluate the scale, quality, and effectiveness of the Federal effort in science and technology."[4]

Here it is a scientist and not a scientific institution who crosses the line between science and politics thereby becoming a political body himself. Not so much scientifically founded political decisions are in focus here – that would be a mission science could hardly manage – but science itself, its function in a science-oriented world with science-supported technology and its own evolution. Compared to the method of employing a scientific institution for scientific policy advice, like it is done in Germany with the *Leopoldina*, would this not be a better solution? Science could remain what it is, namely science, not politics, and politics would not have to engage in the complicated business of fitting scientific know-how into its own structure and decision-finding mechanisms.

But something else also has to be mentioned: even with political indicators, policy advice is not essentially the task of socially responsible science, rather it is *society advice*. In a certain sense, scientific knowledge is not owned by politics, but by society. And not only in "technological" issues does this become evident – when implementing scientific know-how in technical and other products – but in an essential sense also in a socio-scientific and humanities-oriented context. Beside social sciences and humanities, society or (rational) culture itself is the addressee of this knowledge; politics only to the extent of its speaking and acting on behalf of the society. From this perspective, science does not hire itself out to politics with all the consequential problems relevant to its own nature, its own essence. For society and its culture, including political culture, it would rather do the work of *enlightenment*, in a broad sense, including the development and structure of the social and cultural world. Its responsibility would then be directly for this world and not primarily for politics focussing on its own targets.

Strictly speaking, the perspective science uses to accomplish its mission is not really professional or disciplinary, it could better be described as an *inter-* or *transdisciplinary* perspective. And this is not a whim of science but objectively

---

3 Self-description of the Irish Chief Scientific Advisor to the Government. http://www.c-s.ie/ (October 2015).

4 "Office of Science and Technology Policy," *Wikipedia, the free encyclopedia*. http://en.wikipedia.org/wiki/Office_of_Science_and_Technology_Policy (October 2015).

necessary. The problems of society and the world are not going to do us the favour of defining themselves as problems ready to be solved professionally or disciplinarily – look at environment, energy or health problems, for example. That is why in these issues the mission of science (its becoming involved) is always a task that can only be accomplished by drawing on diverse but always relevant scientific knowledge in different fields. Scientific academies, for example, could (and should) be locations where science applies itself to such tasks without being handicapped by institutional fragmentations like subjects and disciplines. Looking at it this way, the political decision for academies in an advising function is logical, as long as the academy understands itself as a working academy with a transdisciplinary working form. Only, it should not be seen as commissioned work in political everyday life, and the actual mission of society advice should not be lost from sight.

Policy advice in idle motion? Only if the conditions referred to here are disregarded and if the required knowledge transfer fails due to two incapacities: science blinding out society and politics ignoring science in its own nature.

20 Science – the Last Adventure

Science is adventurous, and it has an utopian character. Utopias are immigrant wishes; science is a way to recover them – in a double sense. In the first sense, we have those utopias imagined by the opponents in the debate concerning modern achievements in biology and medicine. Genetic technology and reproductive medicine are supposed to be like counters in a department store where shoppers can browse the range of products that interest them: the right eye- and hair-colour, skill at sports or in love, mathematical or artistic genius, maybe a small Mozart or Einstein clone. Here the little man can shop for something really great.

Scientific nonsense. It is science itself that makes its own utopias, and which gives them enough of a foundation to distinguish between realistic utopias and science fiction. Otherwise we would have walked on the moon already in Johannes Kepler's time, and the soul would have become physical in the time of Greek atomism. Furthermore, why should we not work on our (biological) nature? We storm from one technical revolution to the next and still we remain naturally imperfect beings, that can scarcely compete in a natural environment, relying on our immediate resources, with a mouse. Cars are continuously improved technically, so why not also that which sits in them? (In spite of what has been said earlier in chapter 16 on the ideas of transhumanism.)

This brings us to a second meaning of utopia in the scientific context: Science is in its very nature utopian. What does this mean? Is science not real? Is science not an essential part of the modern world? Science is, notwithstanding its great successes, not perfect but something perpetually unfinished – not in the sense of a defect, but as something that belongs to the essence of science, to its peculiar infinite character.[1] If science were completable, that is, if at some point everything that can be explained were in fact explained with scientific methods, if at some point all questions were to be answered that can be posed scientifically, and if at some point everything were mastered that can be mastered with scientific methods, then science itself would be a mere means, an artifact, not a process – at least not in the sense that science constitutes precisely the future potential of a technical and rational culture like ours. It would be as if one took rationality or reason to be something completable, to be something that one could at some point have at one's disposal as a perfected good. But rationality and reason are, on the contrary, never completely satisfiable. They are demands upon thought and action that must be repeatedly posed. At

1 See chapter 5.

https://doi.org/10.1515/9783110596687-020

least in the wake of the European Enlightenment, they are demands whose sense lies not in their complete realization but rather in their "infinite" contribution to our orientation.

So, too, in the case of science. In science we find expressed the "infinite" will of humanity to comprehend our world – and ourselves – and even more to make it our own work. That this in turn is an infinite task lies not in the fact that this task is itself utopian – we already live in a world which is to an ever greater extent the product of scientific and technical understanding – but rather in the fact that there are no scientifically final answers to the question of how our world, insofar as it is (also) our product, should look in the end and how humankind, even with scientific means, should understand itself. Furthermore, science is extremely inventive, not only in its results but also in its questions, and it is inexhaustible, just as understanding and reason are inexhaustible. "Utopia" is an expression of this "infinity" of science or of a scientific culture, and it is the expression of the insight that science, again just like understanding and reason, always has its essence before it. It always lives in the awareness that it is not what it is supposed to be, namely – in the words of the German Idealist Johann Gottlieb Fichte – absolute knowledge. Such knowledge is indeed a pure utopia, but a useful one: it keeps the process of science and the process of knowledge in general in motion.

In other words, there is no subscription to the discovery of the new. Many research programmes are boring, and indeed produce boring results, many goals are revealed to be drains on resources, as destinations without access, as premature promises that are followed only by disappointment. Whoever wants to support science – and that should include everyone, considering the modern world's dependence on science – should be sceptical of promises of quick success. Science is not a product that gives in to the demands of development, production and marketing. This holds true just as well of computer science and fusion research, eminent examples of a desirable orientation towards practical applications, as it does for cancer research. For their successes, on which many of our best researchers are still at work, is accompanied by much sweat and persistence. There is no progress to be made in the latter field without a steadily improving knowledge of genetics and molecular biology, and to think otherwise is to demonstrate one's ignorance. Innovation in science takes time, sometimes a lot of it. And an adventurous mind has to have the corresponding stamina.

Does this mean that research is only something for the old, for those who have aged with their research? Must the young first grow old in order to succeed in research? Experience and the history of science speak against this conclusion. Nothing that stands in the way of insights, even those which open roads to the new, not even inexperience. Furthermore, the old are enamoured perhaps less of the new, as they are of the old, treasured insights that are bound up with them and their work. In fact, the new often means that the old is not only old, but out of date, insufficient, and indeed displaced by new developments. After all, who can really be pleased, even in the context of scientific research, when developments exceed one's own work, and relegate the latter to the shadows of the past. That is why established researchers, as Karl R. Popper repeatedly admonished, remain attached to what already exists, to what has already been proven and may well carry their name, whereas younger researchers see their future in opposition to what is established, that is to say in overcoming it.

Nothing in the concept of (theoretical) limits undermines these conclusions. For let us recall what has been said before (dealing with limits of science): Research is not going to run out of questions, above all not in the case where, following principles of critical reflection in scientific research, we accept the insight that scientific knowledge must as a rule be taken to be imperfect, or incomplete and incorrect, though not in the sense of a defect – such a notion would in fact presuppose an attainable perfection or completeness – but in the sense of an openness of scientific knowledge in principle. Paradoxically formulated, the boundlessness of science, in the sense of an interminable progress of knowledge, lies precisely in the limited character of knowledge, in its finitude and corrigibility.

It is just this that provides opportunity to the young. For they do not depend on their own established results, and they are hungry – hungry for the new. That is why the nurturing of the young is of the greatest importance. If this nurturing is neglected, or insufficiently secured, the ground in which research grows is rendered barren. A tendency to immunize, that is to say to isolate one's own preferred truths against external criticism, sets in. One forgets that research is not a terminus, but a permanent departure to new destinations, and that science might be the last adventure – on an earth grown old.

Furthermore, science is not only hard to reckon, but it is often inconstant and remarkably democratic. It blesses organized researchers at work at their desks and in the lab, after many pencils and much Helium are used up, and it blesses the disorganized as well, for instance while shaving, jogging or making coffee. The new often comes unexpectedly, as an unwanted adventure, the good fortune of the undaunted, of the old as well as of the young. And it is in just this that the proper, lasting challenge of science and scientists lies, namely in their supposed foolishness, which is the handmaiden both of truth and of tomorrow's utility.

Acknowledgements

Some parts of this book are based on already published material. Chapters 1, 2, 3, 7, 12, 14 and 16, in a first version, were presented to the Pontifical Academy of Sciences (*Pontificia Academia Scientiarum*) and published in the respective proceedings (*Acta* 22, 2015, pp. 45–53; *Acta* 19, 2008, pp. 162–172; *Acta* 18, 2006, pp. 3–8; *Scripta Varia* 121, 2013, pp. 97–105; *Acta* 21, 2011, pp. 50–58; *Scripta Varia* 105, 2003, pp. 179–187; *Acta* 20, 2009, pp. 494–503). Other versions of chapters have been published in different places: Chapter 4 (under the title: On the Philosophy of Time) in: *European Review: Interdisciplinary Journal of the Academia Europaea* 9 (2001), pp. 19–29; Chapter 6 (under the title: On Transdisciplinarity) in: *Trames: Journal of the Humanities and Social Sciences* 15 (2011), pp. 329–338; Chapter 9 (under the title: The Concept of Causality in Greek Thought) in: P. Machamer and G. Wolters (eds.), *Thinking about Causes: From Greek Philosophy to Modern Physics*, Pittsburgh: University of Pittsburgh Press 2007, pp. 1–13; Chapter 10 in: G. Hermerén *et al.* (eds.), *Trust and Confidence in Scientific Research*, Stockholm: Kungl. Vitterhets Historie och Antikvitets Akademien 2013, pp. 16–22; Chapter 15 in: *European Review: Interdisciplinary Journal of the Academia Europaea* 4 (1996), pp. 293–300; Chapter 17 in: *Kriterion: Journal of Philosophy* No. 23 (2010), pp. 1–4; Chapter 18 in: *International Journal for Education Law and Policy* 3 (2007), pp. 37–40.

https://doi.org/10.1515/9783110596687-021

Index of Names

Achilles 161
Agassi, J. 24
Aiton, E. J. 168
Ajayan, P. M. 63
Albert, H. 141
Alembert, J. le R. d' 43, 132
Anaxagoras 88–90
Angeli, F. 96
Arago, F. 24
Archimedes 111
Aristotle 29, 36, 38–39, 41, 48, 64, 69, 80–83, 85–87, 89–90, 93–98, 100, 102–103, 113, 115, 147, 155
Assézat, J. 43, 132
Athena 130
Audretsch, J. 92, 103
Augustinus, A. 28, 36
Autrum, H. 131

Bacon, F. 117, 120, 131, 182
Bayertz, K. 171
Bellarmine, R. 102
Bennet, M. R. 73
Bernoulli, Jak. 43, 132
Bernoulli, Joh. 43, 132
Beutler, E. 27
Binet, A. 138
Birnbacher, D. 52, 166, 169–170
Blumenberg, H. 101
Boehme, G. 38
Boerhaave, H. 58
Boethius, A. M. T. S. 94
Bohm, D. 5, 34
Bohr, N. 5, 7, 34
Boltzmann, L. 30
Born, M. 15, 49
Bostrom, N. 171
Boyle, R. 113
Brandt, R. 93
Brecht, B. 155
Breuer, Th. 18
Broad, Ch. D. 20
Broecker, W. S. 13
Broglia, R. A. 63
Bromand, J. 21
Bromley, A. 44, 132
Brown, J. R. 80
Bruno, G. 104, 136–137
Buck, R. S. 134
Buldt, B. 53
Burry, J. 47
Butterfield, J. 35
Butts, R. E. 108, 119

Čapek, M. 21, 28
Carnap, R. 126
Carosella, E. D. 69
Carrier, M. 7, 17–18, 20, 42, 54, 72, 124, 126, 128
Caspar, M. 100, 167
Cavalli-Sforza, L. L. 164
Cavendish, H. 23
Chu, St. 60
Churchland, P. M. 72
Churchland, P. S. 72
Clairaut, A.-C. 43, 132
Cohen, I. B. 106–108, 110, 168
Cohen, R. S. 134
Colli, G. 165
Colodny, R. G. 118
Copernicus, N. 99–104, 109, 136, 182
Cotes, R. 112
Crombie, A. C. 107–108
Currie, G. 134

Dacqué, E. 8
Darwin, Ch. 136–138, 167, 182
Davies, J. W. 108, 119
Denton, G. H. 13
Descartes, R. 69, 85, 109–111, 117–120, 163
Diderot, D. 43, 132
Diels, H. 88, 103
Dijksterhuis, E. J. 84
Dingler, H. 105
Driesch, H. 20

https://doi.org/10.1515/9783110596687-022

Du Bois-Reymond, E. 6, 17
Duhem, P. 10
Duncan, A. M. 168
Durkheim, É. 9
Dux, G. 51
Dyck, W. von 100, 167

Earman, J. 16, 19, 28
Ebbesen, T. W. 63
Eccles, J. C. 21, 74
Eddington, A. S. 25
Edleston, J. 112, 114
Einstein, A. 5, 15, 25–26, 33, 49–50, 80, 84–86, 122, 124–125, 150, 159, 180, 187
Emerson, W. 117
Empedokles 88–89
Eser, A. 51
Euclid 43, 118, 132
Euler, L. 43, 119, 132

Falkenburg, B. 73
Faust 43, 131
Feigl, H. 72
Feldman, M. W. 164
Feyerabend, P. 118–119, 127, 134, 141–142
Feynman, R. P. 62
Fichte, J. G. 71, 166, 188
Fontaine, A. 43, 132
Foscarini, P. A. 102
Fourier, J. B. J. 119
Fraunhofer, J. von 61
Fresnel, A. J. 24
Frisch, Ch. 101
Fritz, K. von 87

Galilei, G. 87, 97, 100, 102–103, 105–106, 113, 115–117, 120, 136–137, 155
Garin, E. 165
Garwin, L. 60–61
Gaye, R. K. 89
Gehlen, A. 164
Gerhardt, G. 170
Gethmann, C. F. 52, 66, 69, 153, 171
Ghirardi, G. C. 63
Gibbons, H. 128
Gierer, A. 54
Goedel, K. 53–55
Goethe, J. W. von 27
Gold, Th. 33
Gravesande, W. s' 110, 117
Grünbaum, A. 28–31
Guyer, P. 70

Habermas, J. 170
Hacker, P. M. St. 73
Hamel, J. 102
Hamilton, W. 116
Hammer, F. 100, 167
Hanson, N. R. 107–108
Hardie, R. P. 89
Harris, J. 111–112, 116
Hartsoeker, N. 112
Haug, G. 176
Hawking, St. W. 33, 126
Heckhausen, H. 75
Hegel, G. W. F. 59, 140
Heidegger, M. 12, 38, 70, 165
Heilinger, J. Chr. 70
Heinekamp, A. 160
Heisenberg, W. 14–16, 34, 53, 80, 85, 123–125
Heraclitus 88
Hermes Trismegistos 82, 85, 173
Hertz, H. 120
Hogrebe, W. 21
Hook, R. 110
Hoppe, E. 133
Horgan, J. 44
Hoyningen-Huene, R. 17
Huebscher, A. 166
Humboldt, W. von 151
Humphreys, P. 42
Husserl, E. 70
Huxley, Th. H. 137
Huygens, Chr. 117

Isham, C. 35
Israel, W. 33

Janich, P. 28, 39, 73, 135, 142
Jonas, H. 168–169

Kampen, N. G. van 19

Kant, I. 12, 38, 52, 55, 69–71, 75, 100, 119, 125–126, 130–131, 154, 160, 165, 169–170
Kass, L. R. 171
Kaufmann, M. 170
Keill, J. 117
Keller, E. F. 122
Kelsey, S. 91
Kepler, J. 102–103, 105–106, 111, 114, 117, 167–168, 170, 187
Kiefer, C. 33, 35
Kitcher, Ph. 52
Kocka, J. 57
Koyré, A. 107–108, 111
Krohn, W. 128
Kuhn, Th. S. 25, 127, 133–135

Lafuma, L. 45
Lagarde, P. de 104
Lagrange, J. 119, 132
Lakatos, I. 127, 134–135
Laplace, P. S. 6, 15, 17–19, 21
Lavoisier, A. L. de 23, 58
Lee, K. J. 93
Leibniz, G. W. 69–70, 91, 111–112, 117, 143, 159–160
Leonardo da Vinci 131, 155–162
Leucippus 88
Lewis, R. 176
Libet, B. 76
Lindley, D. 44
Locke, J. XII, 69, 111, 118
Lorenz, K. 4, 20, 54, 71, 91, 124, 164
Lorenzen, P. 127
Luebbe, H. 137
Lyssenko, T. 137–138

Mach, E. 26, 84, 120
Maclaurin, C. 117
Mainzer, K. 3, 92, 103
Malakoff, D. 60
Marchlewitz, I. 160
Martin, B. 112
Martin, J. 117
Marx, K. 9
Massey, G. J. 42
Maudlin, T. 63
Maupertuis, P. L. M. de 43, 110, 117, 132
Maxwell, J. C. 33, 119, 159
Mendel, G. 137
Merton, P. K. 148
Michel, K. M. 141
Midgley, M. 171
Mittelstrass, J. 4–5, 7, 17, 20, 22, 25, 28, 45, 49–50, 53–54, 57, 69, 71–72, 75, 80, 82, 91–93, 103, 124, 126–128, 135, 139, 142, 160, 164, 169, 179
Moldenhauer, E. 141
Montinari, M. 165
Motte, A. 108
Mozart, W. A. 187
Musschenbroek, P. van 110

Nathan, O. 50, 150
Needham, J. 23
Newton, J. XII, 26, 48, 64, 83–86, 103, 105–120, 131
Nietzsche, F. 163, 165, 168–169
Nobis, H. M. 100
Norden, H. 50, 150
Nowotny, H. 128

Oerstedt, H. Chr. 24
Oldenburg, H. 110
Osiander, A. 100–104
Owen, G. E. L. 97

Pagel, W. 23
Pais, A. 25
Pandora 171
Pardies, G. 109–110
Parmenides 85
Pascal, B. 45
Paul III. (Pope) 100
Pemberton, H. 117
Penrose, R. 33–34
Petreius, J. 101
Pico della Mirandola, G. 165, 167–168, 170–171
Pirotta, M. 96
Pistorius, F. 101
Planck, M. 13, 15, 25, 48, 61, 85
Plato 27, 36, 38–39, 80, 85, 87, 89, 90–94, 96–98, 147, 182
Plessner, H. 37, 51, 164, 166, 168

Podolsky, B. 9
Poincaré, H. 126
Poisson, S. D. 24
Popper, K. R. 7, 9, 19, 21, 26, 74, 126–127, 134–135, 189
Price, H. 28, 33, 35
Prigogine, I. 32
Ptolemy, C. 100, 104

Raff, W.-K. 164
Reichenbach, H. 28, 30, 32
Reid, Th. 116
Reiss, H. 12
Rescher, N. 47–48, 132
Rheticus, G. J. 100–102
Rimini, A. 63
Roebbecke, M. 180
Roentgen, W. C. 22–23
Rohault, J. 110
Roller, D. H. D. 106
Rosen, E. 100, 102
Rosen, N. 9
Roughley, N. 171
Ruetsche, L. 42
Rutherford, E. 23, 174

Scheler, M. 164
Schiller, F. 143
Schofield, R. E. 110
Schopenhauer, A. 75, 166, 169
Schroedinger, E. 13–14, 123
Schulman, L. S. 28–29
Schuster, H. G. 16
Schwarz-Hahn, St. 176
Schwemmer, O. 169
Scott, J. F. 110
Scott, P. 128
Sedley, D. 91
Sexl, R. U. 23–24
Shakespeare, W. 161
Siemens, W. von 131
Silver, L. M. 171
Simon, D. 180
Simplicius 103
Singer, W. 68
Smart, J. J. C. 35, 72
Smuts, J. C. 8
Sneed, J. 127
Snow, Ch. P. 161
Socrates 89–90, 94
Sosoe, L. 170
Spencer, H. 45
Spinner, H. F. 141
Stegmüller, W. 127, 142
Stengers, J. 32
Stephan, A. 21
Sticker, B. 100
Sturm, Th. 93

Tarski, A. 7
Teichler, U. 180
Tetens, H. 49
Thiel, Chr. 7, 82
Thomas Aquinas 96
Toulmin, S. 141
Tourneux, M. 43, 132
Turnbull, H. W. 110

Vlastos, G. 90
Voltaire 117

Wagner, J. J. 82
Weber, M. 147, 150
Weber, T. 63
Weinberg, St. 44
Weinfurter, H. 64
Weingart, P. 128
Weischedel, W. 38, 130, 165
Weizsaecker, C. F. von 62
Westerhijden, D. F. 176
Westman, R. S. 102
Wieland, W. 95–97
Wilberforce, S. 137
Winckelmann, J. 150
Wittgenstein, L. 71
Wollgast, S. 6, 17
Wood, A. W. 70
Worral, J. 134
Wrightsman, B. 102
Wuketits, F. M. 136

Zalta, E. N. 72
Zeh, H.-D. 28–29, 33
Zeilinger, A. 64
Zinner, E. 101

Index of Subjects

accountability, principle of 19
aeon 38, 80, 92
analogy, functional 7–8
analogy, structural 7–8
analysis, principled 95, 97–98
anisotropy 28–33
anomaly 134
apparatus states 18
arche 94–95
Aristotle world 80–83, 85
arrow, cosmological 34–35
arrow, gravitational 34
arrow, quantum mechanical 34
arrow, thermodynamic 33–34
artificiality, natural 164, 166–167, 169–171

bios theoretikos 147–148
black hole 33–34
branch systems 30–31
butterfly effect 16

Cartesian world 85
causa 95
causality 16, 58, 75–76, 85, 87–98
causality, non-local 64
causation, law of 87–88
causation, mental 72
causation, principle of 87
chance, absolute 5
chaos 4–5, 16–18, 32
chaos, deterministic 5, 16
cloning, reproductive 50–52
closure, causal 73, 75
cogito ergo sum 111
Columbus world 156
complexity 3–5, 8, 10, 20
complicatedness 4
concepts, theoretical 126–127
conditions, initial/boundary 31, 33
confirmation 8–10
conjunctio 82
consciousness 55, 68–69, 72–75
constructivism 127
Copenhagen interpretation 5, 34, 63–64, 122, 126
corroboration 127
cosmos 38–39, 45, 91
CPT theorem 34
culture 158–161

derivability 54
determination, characteristic 20
determinism 15–16, 18–19, 21, 73, 79, 123, 125
determinism, neuronal 73
determinism, ontological 21
deterministic 13, 16–17
difference, qualitative 20
dignity 170
disciplinarity 57, 59
diversity, radical 72
dualism, pragmatic 74, 76
duration 39
dynamics, statistical 6
dynamis 94

ego 68–71, 74
ego, transcendental 55, 70
Einstein-Podolsky-Rosen paradox 9
Einstein world 85
electrodynamics 125, 159
electromagnetism 24
emergence 10, 19–21, 124
emergence, strong 10, 20–21, 124
emergence, weak 10, 20, 124
empeiria (Aristotelian) 97
empiricism 118–120, 125
energeia 94
enlightenment 130–131, 142, 150, 185, 188
entelechy 20
entropy 29–31, 33
ethics (of science) 99, 152–154
ethics, rational 170
ethics, universal 154

https://doi.org/10.1515/9783110596687-023

ethos 152–154
Euclidean geometry 43, 132
evolution 8, 79, 136
experience, instrumental 97
experience, phenomenal 97
explanation 20

falsifiability 7, 127
falsification 7, 106, 127
finitism, principle of 54
forma mixti 81
forma mundi 100–101, 103–104
forms, noetic 105–106
freedom (of science) 150–151

genetics 6
geometricization 84
gestalt 37–40

Heisenberg relations 15, 123–124
Heisenberg world 85
heliocentrism 100
Hermes world 82, 85, 173
historicism 127
history, external 134, 136, 138
history, internal 134, 136, 138
history, justified 135
holism, confirmation 9–10
holism, designation 8
holism, meaning 10
holism, methodological 9
holism, ontological 9
holism, semantic 10
homo faber 156, 161, 168, 171
homo sapiens 155, 161, 163, 168, 171
H-theorem 30
hypothesis 9–10, 101–116, 119

idealism 125
ideas, theory of 89, 92
identity theory 72–73
imperative, categorical 52, 144, 170
imperative, scientific 157
incommensurability 133
incompleteness theorem 53–55
incomprehensibility, jargon of 175
indeterminacy, mechanical 159
indeterminacy relations 14
indeterminism 19, 73
individualism, methodological 9
instrumentalism 122–123
interactionism 74
interdisciplinarity 57–59
interpretability 7
IQ-test 138
irreversibility 29–32, 34
I-substance 69

Laplace's demon 6, 15, 17–19, 21
law, causal 13
Leibniz world 159–160
Leonardo world 155, 157–162
limits (of knowledge/science) 16–17, 41–56, 132, 149
Logical Empiricism 71, 126–127

Mach's principle 26, 84
macrocosm 82
materialism, eliminative 72
Maxwell-Hertz equations 121
measurability 18
mechanicism 124
mechanics, Newtonian 19
mechanism 8
method, analytical 115–116
method, synthetical 115–116
metodo compositivo 115–116
metodo risolutivo 115–116
microcosm 82
mind-body problem 54, 74
minima naturalia 81
model 4–5
model, analogue 5, 8
model, scale 5
model, theoretical 5
monad 69, 143, 159
monism 72, 74

nanotechnology 62
naturalism 72–73, 76
naturalness, artificial 166–167, 169–171
natura naturans 79, 166
natura naturata 79, 166
nature 79–86

neo-vitalist 20
newtonianism 84
Newton world 83–85

observer, inner 18
ousia 94
owl 55, 141
ozone layer 12

Pandora's Box 171
paradigm 134
Parmenidean world 85
participation (Plato) 90–91
philosophy, experimental 117
phoenix 55
physicalism 6, 71–72
Planck world 85
platonism 125
platonism, Cambridge 83–84
platonism, Christian 80
Plato world 80, 85
pluralism (of theories) 7, 134
Poisson's white spot 24
posthumanism 171
predictability 5, 12, 14–19, 21, 124
principle, metabiological 8
probabilism 79
probability 15
process, evolutionary 127
progress 7, 48–49, 53, 55, 127, 130–133, 149–150

qualitas media 81
quantum computing 64
quantum cryptography 64
quantum field theory 35
quantum mechanics 5, 13–17, 19, 34–35, 63–64, 85, 121, 123, 159
quantum theory 8–9, 121, 125, 160

rationality 93, 131, 140, 147, 155–156
realism 122–123, 125
realism, epistemological 125
realism, ontological 125
reason, practical 150
reconstruction, rational 134–135
reconstructivism 127
reductability 21
reduction 6–8
reduction, ontological 6
reductionism 6, 8, 72–73, 76
regulae philosophandi 107, 111, 117
relationalism 125, 159
relativity, general theory of 25–26, 33–34, 84–85, 121, 125, 159–160
relativity, special theory of 63, 85, 121, 123
representation 160
research, applied 139, 140
research, basic 139, 140
research, concept of 128
research form (of science) 126
responsibility 140, 143, 150–152, 173
reversibility 31
revolution, scientific 135

Schroedinger's equation 13–14, 123
science, normal 133–134
self 68, 70–71, 125
self-consciousness 54, 68–70, 72–74
sensorium Dei 84
separatio 82
similarity 8
simplicity 44
space, absolute 26, 84
space-time 83–84
spin states 14
statistics, quantum 121
structuralism 127
subject, transcendental 70
substance 93–94
substratum 94
superposition 63
super-scientist 19
supervenient 17
synergetics 32–33
synthetic a priori 126
system, chaotic 6
system, quantum mechanical 6
systema mundi 100

telos 95–96
theoria 93, 97, 148
theory dynamics 126–127

theory explication 126
theory form (of science) 126
theory of everything 35, 125
theory structure 126
thermodynamics, classical 32–33
thermodynamics, irreversible 31–32
thermodynamics, non-equilibrium 32–33
thermodynamics, non-linear 139
thermodynamics, phenomenological 6
thermodynamics, second law of 29–33, 161
time, arrow of 33–35, 39
time, mental 36
transdisciplinarity 57, 59–60, 62, 64–67, 185
transhumanism 171
transmutatio 82
transsubjectivity 151
truthfulness 154
Turing-machine 55
type-identitiy 72–73

uncertainty principle 53
undecidability theorem 53–55
unity (of science) 126
unity of nature 62
universality, causal 13
universals 125
unpredictability 18, 21
utopia 187–188

verification 7, 127
vitalism 8, 124

Wheeler-de Witt equation 35
white hole 34
will, free 13, 74–76
world, deterministic 5–6, 15–16, 21, 123–124
world soul 80, 92
world view (picture) 17, 79, 88